Water Conservation and Management

Water Conservation and Management

Edited by
Cole Gray

Larsen & Keller
www.larsen-keller.com

Water Conservation and Management
Edited by Cole Gray
ISBN: 978-1-63549-288-0 (Hardback)

© 2017 Larsen & Keller

☰ Larsen & Keller

Published by Larsen and Keller Education,
5 Penn Plaza,
19th Floor,
New York, NY 10001, USA

Cataloging-in-Publication Data

Water conservation and management / edited by Cole Gray.
    p. cm.
Includes bibliographical references and index.
ISBN 978-1-63549-288-0
1. Water conservation. 2. Water-supply. 3. Water-supply--Management. 4. Water--Pollution.
5. Water in agriculture. I. Gray, Cole.
TD388 .W38 2017
363.61--dc23

The publisher's policy is to use permanent paper from mills that operate a sustainable forestry policy. Furthermore, the publisher ensures that the text paper and cover boards used have met acceptable environmental accreditation standards.

Printed and bound in the United States of America.

For more information regarding Larsen and Keller Education and its products, please visit the publisher's website www.larsen-keller.com

Table of Contents

Preface

Freshwater is a scarce resource and there is limited distribution of freshwater on the Earth's surface. Therefore, there is a necessity to conserve it for this and upcoming generations. Water conservation refers to the optimum utilization of water and crafting policies, laws, regulations and rules to ensure its proper use. In this book, fundamental methods and techniques that conserve water and mitigate ecological disaster are discussed in detail. The book aims to shed light on some of the unexplored aspects of water conservation. It is compiled in such a manner, that it will provide in-depth knowledge about the theory and practice of this field. The topics covered in it offer the readers new insights about this subject. This textbook aims to serve as a resource guide for students and contribute to the growth of the discipline.

A short introduction to every chapter is written below to provide an overview of the content of the book:

Chapter 1 - To protect water resources and water bodies, and to meet the future demand of fresh water is called water conservation. It includes all the measures and plans to manage fresh water. This chapter provides an integrated study on water conservation and its importance in today's time; **Chapter 2 -** Water resources are sources of water that are useful or possibly useful. The lack of adequate water is water scarcity. This text paraphrases concepts such as water scarcity, water security, saltwater intrusion and wastewater. This chapter is a compilation of the various branches of water resources that form an integral part of the broader subject matter; **Chapter 3 -** The concerns of water conservation elucidated are water scarcity, water security, peak water, wastewater, etc. Water scarcity involves water shortage, water stress and water crisis while wastewater is any water that has been adversely affected by human influence. Conservation of water should be of utmost importance, not only for the present generation but also for the future generation; **Chapter 4 -** The contamination of water bodies is water pollution. The aspects of water pollution dealt within this chapter are marine pollution, ocean acidification, thermal pollution, nutrient pollution and nonpoint source pollution. The topics discussed in the chapter are of great importance to broaden the existing knowledge on water pollution; **Chapter 5 -** This chapter discusses the strategies of water conservation in a critical manner providing key analysis to the subject matter. It essentially explains to the reader approaches such as water resource management, wastewater treatment, rainwater harvesting and water metering. The aspects elucidated in this chapter are of vital importance, and provide a better understanding of water conservation; **Chapter 6 -** This chapter discusses the techniques used for water conservation in agriculture. The techniques discussed within this chapter are irrigation, surface irrigation, drip irrigation and other approaches of water conservation. Due to environmental concerns and the depletion of fresh water resources, it is very important to further develop and practice water conservation.

I extend my sincere thanks to the publisher for considering me worthy of this task. Finally, I thank my family for being a source of support and help.

Editor

Introduction to Water Conservation

To protect water resources and water bodies, and to meet the future demand of fresh water is called water conservation. It includes all the measures and plans to manage fresh water. This chapter provides an integrated study on water conservation and its importance in today's time.

Water conservation encompasses the policies, strategies and activities made to manage fresh water as a sustainable resource, to protect the water environment, and to meet current and future human demand. Population, household size, and growth and affluence all affect how much water is used. Factors such as climate change have increased pressures on natural water resources especially in manufacturing and agricultural irrigation.

United States postal stamp advocating water conservation.

The goals of water conservation efforts include:

- Ensuring availability of water for future generations where the withdrawal of fresh water from an ecosystem does not exceed its natural replacement rate.

- Energy conservation as water pumping, delivery and waste water treatment facilities consume a significant amount of energy. In some regions of the world over 15% of total electricity consumption is devoted to water management.

- Habitat conservation where minimizing human water use helps to preserve freshwater habitats for local wildlife and migrating waterfowl, but also water quality.

Strategies

The key activities that benefit water conservation are as follows :

1. Any beneficial deduction in water loss, use and waste of resources.

2. Avoiding any damage to water quality.

3. Improving water management practices that reduce the use or enhance the beneficial use of water.

Social Solutions

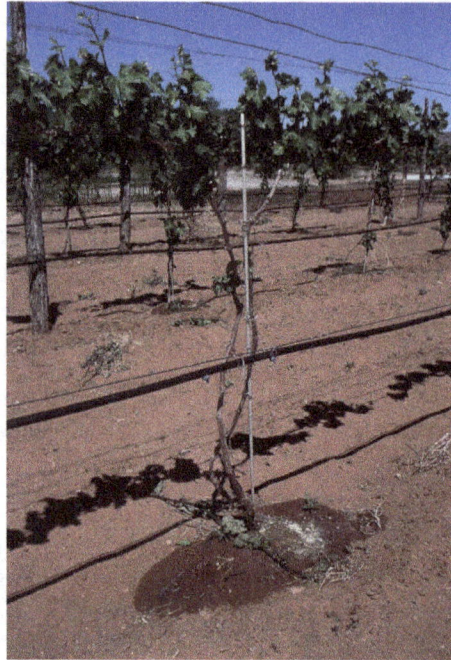

Drip irrigation system in New Mexico

Water conservation programs involved in social solutions are typically initiated at the local level, by either municipal water utilities or regional governments. Common strategies include public outreach campaigns, tiered water rates (charging progressively higher prices as water use increases), or restrictions on outdoor water use such as lawn watering and car washing. Cities in dry climates often require or encourage the installation of xeriscaping or natural landscaping in new homes to reduce outdoor water usage.

One fundamental conservation goal is universal metering. The prevalence of residential water metering varies significantly worldwide. Recent studies have estimated that water supplies are metered in less than 30% of UK households, and about 61% of urban Canadian homes (as of 2001). Although individual water meters have often been considered impractical in homes with private wells or in multifamily buildings, the U.S. Environmental Protection Agency estimates that metering alone can reduce consumption by 20 to 40 percent. In addition to raising consumer awareness of their water use, metering is also an important way to identify and localize water leakage. Water metering would benefit society in the long run it is proven that water metering increases the efficiency of the entire water system, as well as help unnecessary expenses for individuals for years to come. One would be unable to waste water unless they are willing to pay the extra charges, this way the water department would be able to monitor water usage by public, domestic and manufacturing services.

Some researchers have suggested that water conservation efforts should be primarily directed at farmers, in light of the fact that crop irrigation accounts for 70% of the world's fresh water use. The agricultural sector of most countries is important both economically and politically, and water subsidies are common. Conservation advocates have urged removal of all subsidies to force farmers to grow more water-efficient crops and adopt less wasteful irrigation techniques.

New technology poses a few new options for consumers, features such and full flush and half flush when using a toilet are trying to make a difference in water consumption and waste. Also available in our modern world is shower heads that help reduce wasting water, old shower heads are said to use 5-10 gallons per minute. All new fixtures available are said to use 2.5 gallons per minute and offer equal water coverage.

Household Applications

The Home Water Works website contains useful information on household water conservation. Contrary to popular view, experts suggest the most efficient way is replacing toilets and retrofitting washers.

Water-saving technology for the home includes:

1. Low-flow shower heads sometimes called energy-efficient shower heads as they also use less energy

2. Low-flush toilets and composting toilets. These have a dramatic impact in the developed world, as conventional Western toilets use large volumes of water

3. Dual flush toilets created by Caroma includes two buttons or handles to flush different levels of water. Dual flush toilets use up to 67% less water than conventional toilets

4. Faucet aerators, which break water flow into fine droplets to maintain "wetting effectiveness" while using less water. An additional benefit is that they reduce splashing while washing hands and dishes

5. Raw water flushing where toilets use sea water or non-purified water

6. Waste water reuse or recycling systems, allowing:

    o   Reuse of graywater for flushing toilets or watering gardens

    o   Recycling of wastewater through purification at a water treatment plant.

7. Rainwater harvesting

8. High-efficiency clothes washers

9. Weather-based irrigation controllers

10. Garden hose nozzles that shut off water when it is not being used, instead of letting a hose run.

11. Low flow taps in wash basins

12. Swimming pool covers that reduce evaporation and can warm pool water to reduce water, energy and chemical costs.

13. Automatic faucet is a water conservation faucet that eliminates water waste at the faucet. It automates the use of faucets without the use of hands.

Commercial Applications

Many water-saving devices (such as low-flush toilets) that are useful in homes can also be useful for business water saving. Other water-saving technology for businesses includes:

- Waterless urinals

- Waterless car washes

- Infrared or foot-operated taps, which can save water by using short bursts of water for rinsing in a kitchen or bathroom

- Pressurized waterbrooms, which can be used instead of a hose to clean sidewalks

- X-ray film processor re-circulation systems

- Cooling tower conductivity controllers

- Water-saving steam sterilizers, for use in hospitals and health care facilities

- Rain water harvesting

- Water to Water heat exchangers.

Agricultural Applications

Overhead irrigation, center pivot design

For crop irrigation, optimal water efficiency means minimizing losses due to evaporation, runoff or subsurface drainage while maximizing production. An evaporation pan in combination with specific crop correction factors can be used to determine how much water is needed to satisfy plant requirements. Flood irrigation, the oldest and most common type, is often very uneven in distribution, as parts of a field may receive excess water in order to deliver sufficient quantities to other parts. Overhead irrigation, using center-pivot or lateral-moving sprinklers, has the potential for a much more equal and controlled distribution pattern. Drip irrigation is the most expensive and least-used type, but offers the ability to deliver water to plant roots with minimal losses. However, drip irrigation is increasingly affordable, especially for the home gardener and in light of rising water rates. There are also cheap effective methods similar to drip irrigation such as the use of soaking hoses that can even be submerged in the growing medium to eliminate evaporation.

As changing irrigation systems can be a costly undertaking, conservation efforts often concentrate on maximizing the efficiency of the existing system. This may include chiseling compacted soils, creating furrow dikes to prevent runoff, and using soil moisture and rainfall sensors to optimize irrigation schedules. Usually large gains in efficiency are possible through measurement and more effective management of the existing irrigation system. The 2011 UNEP Green Economy Report notes that "[i]mproved soil organic matter from the use of green manures, mulching, and recycling of crop residues and animal manure increases the water holding capacity of soils and their ability to absorb water during torrential rains", which is a way to optimize the use of rainfall and irrigation during dry periods in the season.

References

- Environment Canada (2005). Municipal Water Use, 2001 Statistics (PDF) (Report). Retrieved 2010-02-02. Cat. No. En11-2/2001E-PDF. ISBN 0-662-39504-2. p. 3.

- Hermoso, Virgilio; Abell, Robin; Linke, Simon; Boon, Philip (2016). "The role of protected areas for freshwater biodiversity conservation: challenges and opportunities in a rapidly changing world".

- U.S. Environmental Protection Agency (EPA) (2002). Cases in Water Conservation (PDF) (Report). Retrieved 2010-02-02. Document No. EPA-832-B-02-003.

- Albuquerque Bernalillo County Water Utility Authority (2009-02-06). "Xeriscape Rebates". Albuquerque, NM. Retrieved 2010-02-02.

2

Water Resources: An Overview

Water resources are sources of water that are useful or possibly useful. The lack of adequate water is water scarcity. This text paraphrases concepts such as water scarcity, water security, saltwater intrusion and wastewater. This chapter is a compilation of the various branches of water resources that form an integral part of the broader subject matter.

Water Resources

Water resources are sources of water that are useful or potentially useful. Uses of water include agricultural, industrial, household, recreational and environmental activities. The majority of human uses require fresh water.

Distribution of Earth's Water

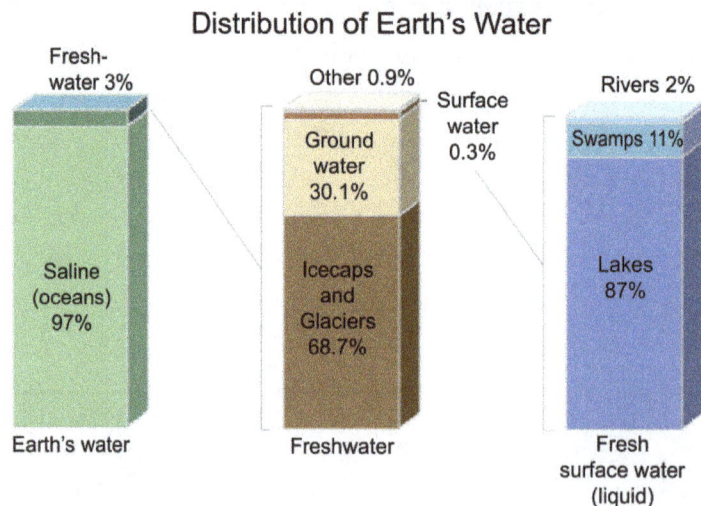

A graphical distribution of the locations of water on Earth. Only 3% of the Earth's water is fresh water. Most of it is in icecaps and glaciers (69%) and groundwater (30%), while all lakes, rivers and swamps combined only account for a small fraction (0.3%) of the Earth's total freshwater reserves.

97% of the water on the Earth is salt water and only three percent is fresh water; slightly over two thirds of this is frozen in glaciers and polar ice caps. The remaining unfrozen freshwater is found mainly as groundwater, with only a small fraction present above ground or in the air.

Fresh water is a renewable resource, yet the world's supply of groundwater is steadily decreasing, with depletion occurring most prominently in Asia and North America, although it is still unclear how much natural renewal balances this usage, and whether ecosystems are threatened. The framework for allocating water resources to water users (where such a framework exists) is known as water rights.

Sources of Fresh Water

Surface Water

Lake Chungará and Parinacota volcano in northern Chile

Surface water is water in a river, lake or fresh water wetland. Surface water is naturally replenished by precipitation and naturally lost through discharge to the oceans, evaporation, evapotranspiration and groundwater recharge.

Although the only natural input to any surface water system is precipitation within its watershed, the total quantity of water in that system at any given time is also dependent on many other factors. These factors include storage capacity in lakes, wetlands and artificial reservoirs, the permeability of the soil beneath these storage bodies, the runoff characteristics of the land in the watershed, the timing of the precipitation and local evaporation rates. All of these factors also affect the proportions of water loss.

Human activities can have a large and sometimes devastating impact on these factors. Humans often increase storage capacity by constructing reservoirs and decrease it by draining wetlands. Humans often increase runoff quantities and velocities by paving areas and channelizing stream flow.

The total quantity of water available at any given time is an important consideration. Some human water users have an intermittent need for water. For example, many farms require large quantities of water in the spring, and no water at all in the winter. To supply such a farm with water, a surface water system may require a large storage capacity to collect water throughout the year and release it in a short period of time. Other users have a continuous need for water, such as a power plant that requires water for cooling. To supply such a power plant with water, a surface water system only needs enough storage capacity to fill in when average stream flow is below the power plant's need.

Nevertheless, over the long term the average rate of precipitation within a watershed is the upper bound for average consumption of natural surface water from that watershed.

Natural surface water can be augmented by importing surface water from another watershed through a canal or pipeline. It can also be artificially augmented from any of the other sources listed here, however in practice the quantities are negligible. Humans can also cause surface water to be "lost" (i.e. become unusable) through pollution.

Brazil is the country estimated to have the largest supply of fresh water in the world, followed by Russia and Canada.

Panorama of a natural wetland (Sinclair Wetlands, New Zealand)

Under River Flow

Throughout the course of a river, the total volume of water transported downstream will often be a combination of the visible free water flow together with a substantial contribution flowing through rocks and sediments that underlie the river and its floodplain called the hyporheic zone. For many rivers in large valleys, this unseen component of flow may greatly exceed the visible flow. The hyporheic zone often forms a dynamic interface between surface water and groundwater from aquifers, exchanging flow between rivers and aquifers that may be fully charged or depleted. This is especially significant in karst areas where pot-holes and underground rivers are common.

Groundwater

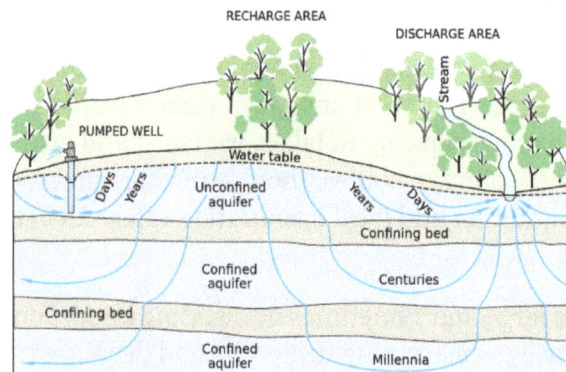

Relative groundwater travel times in the subsurface

Groundwater is fresh water located in the subsurface pore space of soil and rocks. It is also water that is flowing within aquifers below the water table. Sometimes it is useful to make a distinction between groundwater that is closely associated with surface water and deep groundwater in an aquifer (sometimes called "fossil water").

A shipot is a common water source in Central Ukrainian villages

Groundwater can be thought of in the same terms as surface water: inputs, outputs and storage. The critical difference is that due to its slow rate of turnover, groundwater storage is generally much larger (in volume) compared to inputs than it is for surface water. This difference makes it easy for humans to use groundwater unsustainably for a long time without severe consequences. Nevertheless, over the long term the average rate of seepage above a groundwater source is the upper bound for average consumption of water from that source.

The natural input to groundwater is seepage from surface water. The natural outputs from groundwater are springs and seepage to the oceans.

If the surface water source is also subject to substantial evaporation, a groundwater source may become saline. This situation can occur naturally under endorheic bodies of water, or artificially under irrigated farmland. In coastal areas, human use of a groundwater source may cause the direction of seepage to ocean to reverse which can also cause soil salinization. Humans can also cause groundwater to be "lost" (i.e. become unusable) through pollution. Humans can increase the input to a groundwater source by building reservoirs or detention ponds.

Frozen Water

Iceberg near Newfoundland

Several schemes have been proposed to make use of icebergs as a water source, however to date this has only been done for research purposes. Glacier runoff is considered to be surface water.

The Himalayas, which are often called "The Roof of the World", contain some of the most extensive and rough high altitude areas on Earth as well as the greatest area of glaciers and permafrost outside of the poles. Ten of Asia's largest rivers flow from there, and more than a billion people's livelihoods depend on them. To complicate matters, temperatures there are rising more rapidly than the global average. In Nepal, the temperature has risen by 0.6 degrees Celsius over the last decade, whereas globally, the Earth has warmed approximately 0.7 degrees Celsius over the last hundred years.

Desalination

Desalination is an artificial process by which saline water (generally sea water) is converted to fresh water. The most common desalination processes are distillation and reverse osmosis. Desalination is currently expensive compared to most alternative sources of water, and only a very small fraction of total human use is satisfied by desalination. It is only economically practical for high-valued uses (such as household and industrial uses) in arid areas. The most extensive use is in the Persian Gulf.

Water Uses

Agricultural

It is estimated that 70% of worldwide water is used for irrigation, with 15-35% of irrigation withdrawals being unsustainable. It takes around 2,000 - 3,000 litres of water to produce enough food to satisfy one person's daily dietary need. This is a considerable amount, when compared to that required for drinking, which is between two and five litres. To produce food for the now over 7 billion people who inhabit the planet today requires the water that would fill a canal ten metres deep, 100 metres wide and 2100 kilometres long.

Increasing Water Scarcity

Around fifty years ago, the common perception was that water was an infinite resource. At that time, there were fewer than half the current number of people on the planet. People were not as wealthy as today, consumed fewer calories and ate less meat, so less water was needed to produce their food. They required a third of the volume of water we presently take from rivers. Today, the competition for water resources is much more intense. This is because there are now seven billion people on the planet, their consumption of water-thirsty meat and vegetables is rising, and there is increasing competition for water from industry, urbanisation biofuel crops, and water reliant food items. In the future, even more water will be needed to produce food because the Earth's population is forecast to rise to 9 billion by 2050. An additional 2.5 or 3 billion people, choosing to eat fewer cereals and more meat and vegetables could add an additional five million kilometres to the virtual canal mentioned above.

An assessment of water management in agriculture sector was conducted in 2007 by the International Water Management Institute in Sri Lanka to see if the world had sufficient water to provide food for its growing population. It assessed the current availability of water for agriculture on a global scale and mapped out locations suffering from water scarcity. It found that a fifth of the world's people, more than 1.2 billion, live in areas of physical water scarcity, where there is not enough water to meet all demands. One third of the worlds population does not have access to clean drinking water, which is more than 2.3 billion people. A further 1.6 billion people live in areas experiencing economic water scarcity, where the lack of investment in water or insufficient human capacity make it impossible for authorities to satisfy the demand for water. The report found that it would be possible to produce the food required in future, but that continuation of today's food production and environmental trends would lead to crises in many parts of the world. To avoid a global water crisis, farmers will have to strive to increase productivity to meet growing demands for food, while industry and cities find ways to use water more efficiently.

In some areas of the world, irrigation is necessary to grow any crop at all, in other areas it permits more profitable crops to be grown or enhances crop yield. Various irrigation methods involve different trade-offs between crop yield, water consumption and capital cost of equipment and structures. Irrigation methods such as furrow and overhead sprinkler irrigation are usually less expensive but are also typically less efficient, because much of the water evaporates, runs off or drains below the root zone. Other irrigation methods considered to be more efficient include drip or trickle irrigation, surge irrigation, and some types of sprinkler systems where the sprinklers are operated near ground level. These types of systems, while more expensive, usually offer greater potential to minimize runoff, drainage and evaporation. Any system that is improperly managed can be wasteful, all methods have the potential for high efficiencies under suitable conditions, appropriate irrigation timing and management. Some issues that are often insufficiently considered are salinization of groundwater and contaminant accumulation leading to water quality declines.

As global populations grow, and as demand for food increases in a world with a fixed water supply, there are efforts under way to learn how to produce more food with less water, through improvements in irrigation methods and technologies, agricultural water management, crop types, and water monitoring. Aquaculture is a small but growing agricultural use of water. Freshwater commercial fisheries may also be considered as agricultural uses of water, but have generally been assigned a lower priority than irrigation.

Industrial

A power plant in Poland

It is estimated that 22% of worldwide water is used in industry. Major industrial users include hydroelectric dams, thermoelectric power plants, which use water for cooling, ore and oil refineries, which use water in chemical processes, and manufacturing plants, which use water as a solvent. Water withdrawal can be very high for certain industries, but consumption is generally much lower than that of agriculture.

Water is used in renewable power generation. Hydroelectric power derives energy from the force of water flowing downhill, driving a turbine connected to a generator. This hydroelectricity is a low-cost, non-polluting, renewable energy source. Significantly, hydroelectric power can also be used for load following unlike most renewable energy sources which are intermittent. Ultimately, the energy in a hydroelectric powerplant is supplied by the sun. Heat from the sun evaporates water, which condenses as rain in higher altitudes and flows downhill. Pumped-storage hydroelectric plants also exist, which use grid electricity to pump water uphill when demand is low, and use the stored water to produce electricity when demand is high.

Hydroelectric power plants generally require the creation of a large artificial lake. Evaporation from this lake is higher than evaporation from a river due to the larger surface area exposed to the elements, resulting in much higher water consumption. The process of driving water through the turbine and tunnels or pipes also briefly removes this water from the natural environment, creating water withdrawal. The impact of this withdrawal on wildlife varies greatly depending on the design of the powerplant.

Pressurized water is used in water blasting and water jet cutters. Also, very high pressure water guns are used for precise cutting. It works very well, is relatively safe, and is not harmful to the environment. It is also used in the cooling of machinery to prevent overheating, or prevent saw blades from overheating. This is generally a very small source of water consumption relative to other uses.

Water is also used in many large scale industrial processes, such as thermoelectric power production, oil refining, fertilizer production and other chemical plant use, and natural gas extraction from shale rock. Discharge of untreated water from industrial uses is pollution. Pollution includes discharged solutes (chemical pollution) and increased water temperature (thermal pollution). Industry requires pure water for many applications and utilizes a variety of purification techniques both in water supply and discharge. Most of this pure water is generated on site, either from natural freshwater or from municipal grey water. Industrial consumption of water is generally much lower than withdrawal, due to laws requiring industrial grey water to be treated and returned to the environment. Thermoelectric powerplants using cooling towers have high consumption, nearly equal to their withdrawal, as most of the withdrawn water is evaporated as part of the cooling process. The withdrawal, however, is lower than in once-through cooling systems.

Domestic

Drinking water

It is estimated that 8% of worldwide water use is for domestic purposes. These include drinking water, bathing, cooking, toilet flushing, cleaning, laundry and gardening. Basic domestic water requirements have been estimated by Peter Gleick at around 50 liters per person per day, excluding water for gardens. Drinking water is water that is of sufficiently high quality so that it can be consumed or used without risk of immediate or long term harm. Such water is commonly called potable water. In most developed countries, the water supplied to domestic, commerce and industry is all of drinking water standard even though only a very small proportion is actually consumed or used in food preparation.

Recreation

Whitewater rapids

Recreational water use is usually a very small but growing percentage of total water use. Recreational water use is mostly tied to reservoirs. If a reservoir is kept fuller than it would otherwise be for recreation, then the water retained could be categorized as recreational usage. Release of water from a few reservoirs is also timed to enhance whitewater boating, which also could be considered a recreational usage. Other examples are anglers, water skiers, nature enthusiasts and swimmers.

Recreational usage is usually non-consumptive. Golf courses are often targeted as using excessive amounts of water, especially in drier regions. It is, however, unclear whether recreational irrigation (which would include private gardens) has a noticeable effect on water resources. This is largely due to the unavailability of reliable data. Additionally, many golf courses utilize either primarily or exclusively treated effluent water, which has little impact on potable water availability.

Some governments, including the Californian Government, have labelled golf course usage as agricultural in order to deflect environmentalists' charges of wasting water. However, using the above figures as a basis, the actual statistical effect of this reassignment is close to zero. In Arizona, an organized lobby has been established in the form of the Golf Industry Association, a group focused on educating the public on how golf impacts the environment.

Recreational usage may reduce the availability of water for other users at specific times and places. For example, water retained in a reservoir to allow boating in the late summer is not available to farmers during the spring planting season. Water released for whitewater rafting may not be available for hydroelectric generation during the time of peak electrical demand.

Environmental

Explicit environment water use is also a very small but growing percentage of total water use. Environmental water may include water stored in impoundments and released for environmental purposes (held environmental water), but more often is water retained in waterways through regulatory limits of abstraction. Environmental water usage includes watering of natural or artificial wetlands, artificial lakes intended to create wildlife habitat, fish ladders, and water releases from reservoirs timed to help fish spawn, or to restore more natural flow regimes

Like recreational usage, environmental usage is non-consumptive but may reduce the availability of water for other users at specific times and places. For example, water release from a reservoir to help fish spawn may not be available to farms upstream, and water retained in a river to maintain waterway health would not be available to water abstractors downstream.

Water Stress

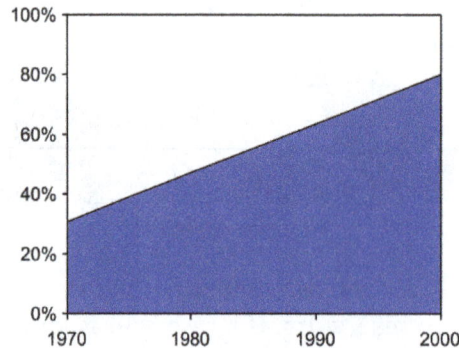

Estimate of the share of people in developing countries with access to drinking water 1970–2000

The concept of water stress is relatively simple: According to the World Business Council for Sustainable Development, it applies to situations where there is not enough water for all uses, whether agricultural, industrial or domestic. Defining thresholds for stress in terms of available water per capita is more complex, however, entailing assumptions about water use and its efficiency. Nevertheless, it has been proposed that when annual per capita renewable freshwater availability is less than 1,700 cubic meters, countries begin to experience periodic or regular water stress. Below 1,000 cubic meters, water scarcity begins to hamper economic development and human health and well-being.

Population Growth

In 2000, the world population was 6.2 billion. The UN estimates that by 2050 there will be an additional 3.5 billion people with most of the growth in developing countries that already suffer water stress. Thus, water demand will increase unless there are corresponding increases in water conservation and recycling of this vital resource. In building on the data presented here by the UN, the World Bank goes on to explain that access to water for producing food will be one of the main challenges in the decades to come. Access to water will need to be balanced with the importance of managing water itself in a sustainable way while taking into account the impact of climate change, and other environmental and social variables.

Expansion of Business Activity

Business activity ranging from industrialization to services such as tourism and entertainment continues to expand rapidly. This expansion requires increased water services including both supply and sanitation, which can lead to more pressure on water resources and natural ecosystem.

Rapid Urbanization

The trend towards urbanization is accelerating. Small private wells and septic tanks that work well in low-density communities are not feasible within high-density urban areas. Urbanization requires significant investment in water infrastructure in order to deliver water to individuals and to process the concentrations of wastewater – both from individuals and from business. These polluted and contaminated waters must be treated or they pose unacceptable public health risks.

In 60% of European cities with more than 100,000 people, groundwater is being used at a faster rate than it can be replenished. Even if some water remains available, it costs increasingly more to capture it.

Climate Change

Climate change could have significant impacts on water resources around the world because of the close connections between the climate and hydrological cycle. Rising temperatures will increase evaporation and lead to increases in precipitation, though there will be regional variations in rainfall. Both droughts and floods may become more frequent in different regions at different times, and dramatic changes in snowfall and snow melt are expected in mountainous areas. Higher temperatures will also affect water quality in ways that are not well understood. Possible impacts include increased eutrophication. Climate change could also mean an increase in demand for farm irrigation, garden sprinklers, and perhaps even swimming pools. There is now ample evidence that increased hydrologic variability and change in climate has and will continue have a profound impact on the water sector through the hydrologic cycle, water availability, water demand, and water allocation at the global, regional, basin, and local levels.

Depletion of Aquifers

Due to the expanding human population, competition for water is growing such that many of the world's major aquifers are becoming depleted. This is due both for direct human consumption as well as agricultural irrigation by groundwater. Millions of pumps of all sizes are currently extracting groundwater throughout the world. Irrigation in dry areas such as northern China, Nepal and India is supplied by groundwater, and is being extracted at an unsustainable rate. Cities that have experienced aquifer drops between 10 and 50 meters include Mexico City, Bangkok, Beijing, Madras and Shanghai.

Pollution and Water Protection

Water pollution is one of the main concerns of the world today. The governments of numerous countries have striven to find solutions to reduce this problem. Many pollutants threaten water supplies, but the most widespread, especially in developing countries, is the discharge of raw sewage into natural waters; this method of sewage disposal is the most common method in underdeveloped countries, but also is prevalent in quasi-developed countries such as China, India, Nepal and Iran. Sewage, sludge, garbage, and even toxic pollutants are all dumped into the water. Even if sewage is treated, problems still arise. Treated sewage forms sludge, which may be placed in landfills, spread out on land, incinerated or dumped at sea. In addition to sewage, nonpoint source pollution such as agricultural runoff is a significant source of pollution in some parts of the world, along with urban stormwater runoff and chemical wastes dumped by industries and governments.

Polluted water

Water and Conflicts

Competition for water has widely increased, and it has become more difficult to conciliate the necessities for water supply for human consumption, food production, ecosystems and other uses. Water administration is frequently involved in contradictory and complex problems. Approximately 10% of the worldwide annual runoff is used for human necessities. Several areas of the world are flooded, while others have such low precipitations that human life is almost impossible. As population and development increase, raising water demand, the possibility of problems inside a certain country or region increases, as it happens with others outside the region.

Over the past 25 years, politicians, academics and journalists have frequently predicted that disputes over water would be a source of future wars. Commonly cited quotes include: that of former Egyptian Foreign Minister and former Secretary-General of the United Nations Boutrous Ghali, who forecast, "The next war in the Middle East will be fought over water, not politics"; his successor at the UN, Kofi Annan, who in 2001 said, "Fierce competition for fresh water may well become a source of conflict and wars in the future," and the former Vice President of the World Bank, Ismail Serageldin, who said the wars of the next century will be over water unless significant changes in governance occurred. The water wars hypothesis had its roots in earlier research carried out on a small number of transboundary rivers such as the Indus, Jordan and Nile. These particular rivers became the focus because they had experienced water-related disputes. Specific events cited as evidence include Israel's bombing of Syria's attempts to divert the Jordan's headwaters, and military threats by Egypt against any country building dams in the upstream waters of the Nile. However, while some links made between conflict and water were valid, they did not necessarily represent the norm.

The only known example of an actual inter-state conflict over water took place between 2500 and 2350 BC between the Sumerian states of Lagash and Umma. Water stress has most often led to conflicts at local and regional levels. Tensions arise most often within national borders, in the downstream areas of distressed river basins. Areas such as the lower regions of China's Yellow River or the Chao Phraya River in Thailand, for example, have already been experiencing water stress for several years. Water stress can also exacerbate conflicts and political tensions which are not directly caused by water. Gradual reductions over time in the quality and/or quantity of fresh water can add to the instability of a region by depleting the health of a population, obstructing economic development, and exacerbating larger conflicts.

Shared Water Resources can Promote Collaboration

Water resources that span international boundaries are more likely to be a source of collaboration and cooperation than war. Scientists working at the International Water Management Institute have been investigating the evidence behind water war predictions. Their findings show that, while it is true there has been conflict related to water in a handful of international basins, in the rest of the world's approximately 300 shared basins the record has been largely positive. This is exemplified by the hundreds of treaties in place guiding equitable water use between nations sharing water resources. The institutions created by these agreements can, in fact, be one of the most important factors in ensuring cooperation rather than conflict.

The International Union for the Conservation of Nature (IUCN) published the book *Share: Managing water across boundaries*. One chapter covers the functions of trans-boundary institutions and how they can be designed to promote cooperation, overcome initial disputes and find ways of coping with the uncertainty created by climate change. It also covers how the effectiveness of such institutions can be monitored.

Water Shortages

In 2025, water shortages will be more prevalent among poorer countries where resources are limited and population growth is rapid, such as the Middle East, Africa, and parts of Asia. By 2025, large urban and peri-urban areas will require new infrastructure to provide safe water and adequate sanitation. This suggests growing conflicts with agricultural water users, who currently consume the majority of the water used by humans.

Generally speaking the more developed countries of North America, Europe and Russia will not see a serious threat to water supply by the year 2025, not only because of their relative wealth, but more importantly their populations will be better aligned with available water resources. North Africa, the Middle East, South Africa and northern China will face very severe water shortages due to physical scarcity and a condition of overpopulation relative to their carrying capacity with respect to water supply. Most of South America, Sub-Saharan Africa, Southern China and India will face water supply shortages by 2025; for these latter regions the causes of scarcity will be economic constraints to developing safe drinking water, as well as excessive population growth.

Economic Considerations

Water supply and sanitation require a huge amount of capital investment in infrastructure such as pipe networks, pumping stations and water treatment works. It is estimated that Organisation for Economic Co-operation and Development (OECD) nations need to invest at least USD 200 billion per year to replace aging water infrastructure to guarantee supply, reduce leakage rates and protect water quality.

International attention has focused upon the needs of the developing countries. To meet the Millennium Development Goals targets of halving the proportion of the population lacking access to safe drinking water and basic sanitation by 2015, current annual investment on the order of USD 10 to USD 15 billion would need to be roughly doubled. This does not include investments required for the maintenance of existing infrastructure.

Once infrastructure is in place, operating water supply and sanitation systems entails significant ongoing costs to cover personnel, energy, chemicals, maintenance and other expenses. The sources of money to meet these capital and operational costs are essentially either user fees, public funds or some combination of the two. An increasing dimension to consider is the flexibility of the water supply system.

Water Cycle

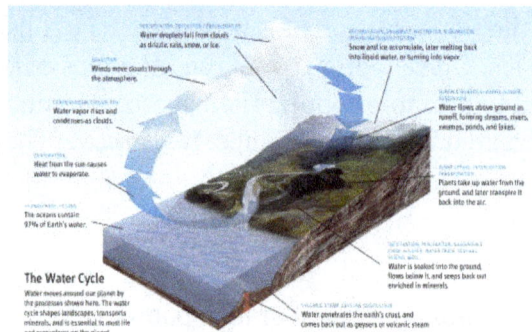

Diagram of the Water Cycle

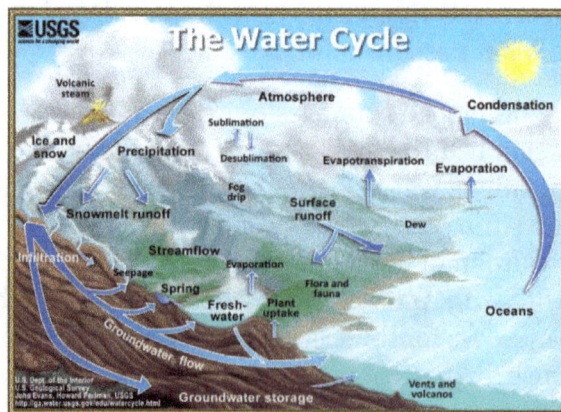

The Water Cycle

The water cycle, also known as the hydrological cycle or the H_2O cycle, describes the continuous movement of water on, above and below the surface of the Earth. The mass of water on Earth remains fairly constant over time but the partitioning of the water into the major reservoirs of ice, fresh water, saline water and atmospheric water is variable depending on a wide range of climatic variables. The water moves from one reservoir to another, such as from river to ocean, or from the ocean to the atmosphere, by the physical processes of evaporation, condensation, precipitation, infiltration, surface runoff, and subsurface flow. In doing so, the water goes through different phases: liquid, solid (ice) and vapor.

The water cycle involves the exchange of energy, which leads to temperature changes. For instance, when water evaporates, it takes up energy from its surroundings and cools the environment. When it condenses, it releases energy and warms the environment. These heat exchanges influence climate.

The evaporative phase of the cycle purifies water which then replenishes the land with freshwater. The flow of liquid water and ice transports minerals across the globe. It is also involved in reshaping the geological features of the Earth, through processes including erosion and sedimentation. The water cycle is also essential for the maintenance of most life and ecosystems on the planet.

Description

The sun, which drives the water cycle, heats water in oceans and seas. Water evaporates as water vapor into the air. Ice and snow can sublimate directly into water vapour. Evapotranspiration is water transpired from plants and evaporated from the soil. The water vapour molecule H_2O, has less density compared to the major components of the atmosphere, nitrogen and oxygen, N_2 and O_2. Due to the significant difference in molecular mass, water vapor in gas form gain height in open air as a result of buoyancy. However, as altitude increases, air pressure decreases and the temperature drops. The lowered temperature causes water vapour to condense into a tiny liquid water droplet which is heavier than the air, such that it falls unless supported by an updraft. A huge concentration of these droplets over a large space up in the atmosphere become visible as cloud. Fog is formed if the water vapour condense near ground level, as a result of moist air and cool air collision or an abrupt reduction in air pressure. Air currents move water vapour around the globe, cloud particles collide, grow, and fall out of the upper atmospheric layers as precipitation. Some precipitation falls as snow or hail, sleet, and can accumulate as ice caps and glaciers, which can store frozen water for thousands of years. Most water falls back into the oceans or onto land as rain, where the water flows over the ground as surface runoff. A portion of runoff enters rivers in valleys in the landscape, with streamflow moving water towards the oceans. Runoff and water emerging from the ground (groundwater) may be stored as freshwater in lakes. Not all runoff flows into rivers, much of it soaks into the ground as infiltration. Some water infiltrates deep into the ground and replenishes aquifers, which can store freshwater for long periods of time. Some infiltration stays close to the land surface and can seep back into surface-water bodies (and the ocean) as groundwater discharge. Some groundwater finds openings in the land surface and comes out as freshwater springs. In river valleys and flood-plains there is often continuous water exchange between surface water and ground water in the hyporheic zone. Over time, the water returns to the ocean, to continue the water cycle.

Processes

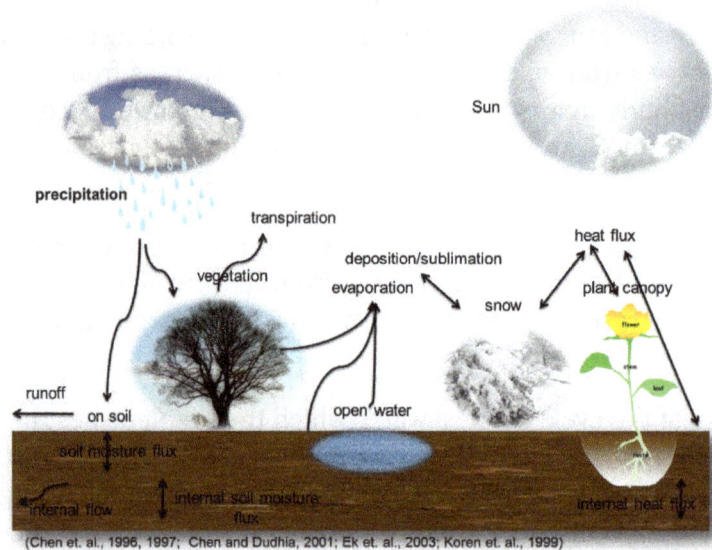

(Chen et. al., 1996, 1997; Chen and Dudhia, 2001; Ek et. al., 2003; Koren et. al., 1999)

Many different processes lead to movements and phase changes in water

Precipitation

Condensed water vapor that falls to the Earth's surface . Most precipitation occurs as rain, but also includes snow, hail, fog drip, graupel, and sleet. Approximately 505,000 km³ (121,000 cu mi) of water falls as precipitation each year, 398,000 km³ (95,000 cu mi) of it over the oceans. The rain on land contains 107,000 km³ (26,000 cu mi) of water per year and a snowing only 1,000 km³ (240 cu mi). 78% of global precipitation occurs over the ocean.

Canopy Interception

The precipitation that is intercepted by plant foliage, eventually evaporates back to the atmosphere rather than falling to the ground.

Snowmelt

The runoff produced by melting snow.

Runoff

The variety of ways by which water moves across the land. This includes both surface runoff and channel runoff. As it flows, the water may seep into the ground, evaporate into the air, become stored in lakes or reservoirs, or be extracted for agricultural or other human uses.

Infiltration

The flow of water from the ground surface into the ground. Once infiltrated, the water becomes soil moisture or groundwater. A recent global study using water stable isotopes, however, shows that not all soil moisture is equally available for groundwater recharge or for plant transpiration.

Subsurface Flow

The flow of water underground, in the vadose zone and aquifers. Subsurface water may return to the surface (e.g. as a spring or by being pumped) or eventually seep into the oceans. Water returns to the land surface at lower elevation than where it infiltrated, under the force of gravity or gravity induced pressures. Groundwater tends to move slowly, and is replenished slowly, so it can remain in aquifers for thousands of years.

Evaporation

The transformation of water from liquid to gas phases as it moves from the ground or bodies of water into the overlying atmosphere. The source of energy for evaporation is primarily solar radiation. Evaporation often implicitly includes transpiration from plants, though together they are specifically referred to as evapotranspiration. Total annual evapotranspiration amounts to approximately 505,000 km³ (121,000 cu mi) of water, 434,000 km³ (104,000 cu mi) of which evaporates from the oceans. 86% of global evaporation occurs over the ocean.

Sublimation

The state change directly from solid water (snow or ice) to water vapor.

Deposition

This refers to changing of water vapor directly to ice.

Advection

The movement of water — in solid, liquid, or vapor states — through the atmosphere. Without advection, water that evaporated over the oceans could not precipitate over land.

Condensation

The transformation of water vapor to liquid water droplets in the air, creating clouds and fog.

Transpiration

The release of water vapor from plants and soil into the air. Water vapor is a gas that cannot be seen.

Percolation

Water flows vertically through the soil and rocks under the influence of gravity

Plate Tectonics

Water enters the mantle via subduction of oceanic crust. Water returns to the surface via volcanism.

Water cycle thus involves many of the intermediate processes.

Residence Times

Average reservoir residence times	
Reservoir	**Average residence time**
Antarctica	20,000 years
Oceans	3,200 years
Glaciers	20 to 100 years
Seasonal snow cover	2 to 6 months
Soil moisture	1 to 2 months
Groundwater: shallow	100 to 200 years
Groundwater: deep	10,000 years
Lakes (see lake retention time)	50 to 100 years
Rivers	2 to 6 months
Atmosphere	9 days

The *residence time* of a reservoir within the hydrologic cycle is the average time a water molecule will spend in that reservoir. It is a measure of the average age of the water in that reservoir.

Groundwater can spend over 10,000 years beneath Earth's surface before leaving. Particularly old groundwater is called fossil water. Water stored in the soil remains there very briefly, because it is spread thinly across the Earth, and is readily lost by evaporation, transpiration, stream flow, or groundwater recharge. After evaporating, the residence time in the atmosphere is about 9 days before condensing and falling to the Earth as precipitation.

The major ice sheets - Antarctica and Greenland - store ice for very long periods. Ice from Antarctica has been reliably dated to 800,000 years before present, though the average residence time is shorter.

In hydrology, residence times can be estimated in two ways. The more common method relies on the principle of conservation of mass and assumes the amount of water in a given reservoir is roughly constant. With this method, residence times are estimated by dividing the volume of the reservoir by the rate by which water either enters or exits the reservoir. Conceptually, this is equivalent to timing how long it would take the reservoir to become filled from empty if no water were to leave (or how long it would take the reservoir to empty from full if no water were to enter).

An alternative method to estimate residence times, which is gaining in popularity for dating groundwater, is the use of isotopic techniques. This is done in the subfield of isotope hydrology.

Changes Over Time

The water cycle describes the processes that drive the movement of water throughout the hydrosphere. However, much more water is "in storage" for long periods of time than is actually moving through the cycle. The storehouses for the vast majority of all water on Earth are the oceans. It is estimated that of the 332,500,000 mi³ (1,386,000,000 km³) of the world's water supply, about

321,000,000 mi³ (1,338,000,000 km³) is stored in oceans, or about 97%. It is also estimated that the oceans supply about 90% of the evaporated water that goes into the water cycle.

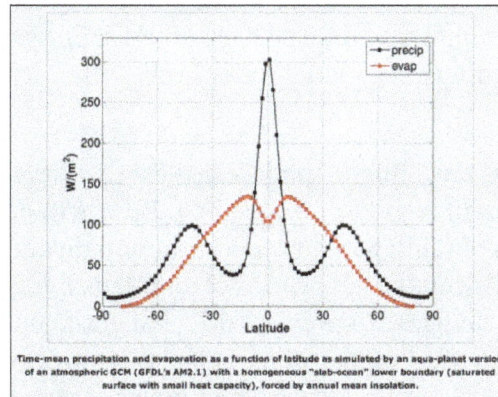

Time-mean precipitation and evaporation as a function of latitude as simulated by an aqua-planet version of an atmospheric GCM (GFDL's AM2.1) with a homogeneous "slab-ocean" lower boundary (saturated surface with small heat capacity), forced by annual mean insolation.

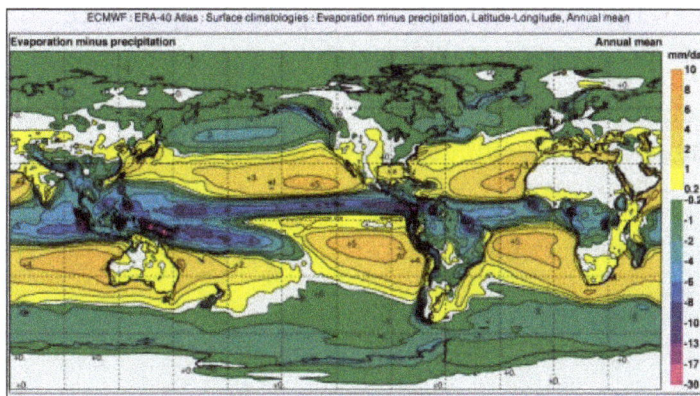

Global map of annual mean evaporation minus precipitation by latitude-longitude

During colder climatic periods more ice caps and glaciers form, and enough of the global water supply accumulates as ice to lessen the amounts in other parts of the water cycle. The reverse is true during warm periods. During the last ice age glaciers covered almost one-third of Earth's land mass, with the result being that the oceans were about 400 ft (122 m) lower than today. During the last global "warm spell," about 125,000 years ago, the seas were about 18 ft (5.5 m) higher than they are now. About three million years ago the oceans could have been up to 165 ft (50 m) higher.

The scientific consensus expressed in the 2007 Intergovernmental Panel on Climate Change (IPCC) Summary for Policymakers is for the water cycle to continue to intensify throughout the 21st century, though this does not mean that precipitation will increase in all regions. In subtropical land areas — places that are already relatively dry — precipitation is projected to decrease during the 21st century, increasing the probability of drought. The drying is projected to be strongest near the poleward margins of the subtropics (for example, the Mediterranean Basin, South Africa, southern Australia, and the Southwestern United States). Annual precipitation amounts are expected to increase in near-equatorial regions that tend to be wet in the present climate, and also at high latitudes. These large-scale patterns are present in nearly all of the climate model simulations conducted at several international research centers as part of the 4th Assessment of the IPCC.

There is now ample evidence that increased hydrologic variability and change in climate has and will continue to have a profound impact on the water sector through the hydrologic cycle, water availability, water demand, and water allocation at the global, regional, basin, and local levels. Research published in 2012 in *Science* based on surface ocean salinity over the period 1950 to 2000 confirm this projection of an intensified global water cycle with salty areas becoming more saline and fresher areas becoming more fresh over the period:

Fundamental thermodynamics and climate models suggest that dry regions will become drier and wet regions will become wetter in response to warming. Efforts to detect this long-term response in sparse surface observations of rainfall and evaporation remain ambiguous. We show that ocean salinity patterns express an identifiable fingerprint of an intensifying water cycle. Our 50-year observed global surface salinity changes, combined with changes from global climate models, present robust evidence of an intensified global water cycle at a rate of 8 ± 5% per degree of surface warming. This rate is double the response projected by current-generation climate models and suggests that a substantial (16 to 24%) intensification of the global water cycle will occur in a future 2° to 3° warmer world.

An instrument carried by the SAC-D satellite launched in June, 2011 measures global sea surface salinity but data collection began only in June, 2011.

Glacial retreat is also an example of a changing water cycle, where the supply of water to glaciers from precipitation cannot keep up with the loss of water from melting and sublimation. Glacial retreat since 1850 has been extensive.

Human activities that alter the water cycle include:

- agriculture
- industry
- alteration of the chemical composition of the atmosphere
- construction of dams
- deforestation and afforestation
- removal of groundwater from wells
- water abstraction from rivers
- urbanization

Effects on Climate

The water cycle is powered from solar energy. 86% of the global evaporation occurs from the oceans, reducing their temperature by evaporative cooling. Without the cooling, the effect of evaporation on the greenhouse effect would lead to a much higher surface temperature of 67 °C (153 °F), and a warmer planet.

Aquifer drawdown or overdrafting and the pumping of fossil water increases the total amount of water in the hydrosphere, and has been postulated to be a contributor to sea-level rise.

Effects on Biogeochemical Cycling

While the water cycle is itself a biogeochemical cycle, flow of water over and beneath the Earth is a key component of the cycling of other biogeochemicals. Runoff is responsible for almost all of the transport of eroded sediment and phosphorus from land to waterbodies. The salinity of the oceans is derived from erosion and transport of dissolved salts from the land. Cultural eutrophication of lakes is primarily due to phosphorus, applied in excess to agricultural fields in fertilizers, and then transported overland and down rivers. Both runoff and groundwater flow play significant roles in transporting nitrogen from the land to waterbodies. The dead zone at the outlet of the Mississippi River is a consequence of nitrates from fertilizer being carried off agricultural fields and funnelled down the river system to the Gulf of Mexico. Runoff also plays a part in the carbon cycle, again through the transport of eroded rock and soil.

Slow Loss Over Geologic Time

The hydrodynamic wind within the upper portion of a planet's atmosphere allows light chemical elements such as Hydrogen to move up to the exobase, the lower limit of the exosphere, where the gases can then reach escape velocity, entering outer space without impacting other particles of gas. This type of gas loss from a planet into space is known as planetary wind. Planets with hot lower atmospheres could result in humid upper atmospheres that accelerate the loss of hydrogen.

History of Hydrologic Cycle Theory

Floating Land Mass

In ancient times, it was thought that the land mass floated on a body of water, and that most of the water in rivers has its origin under the earth. Examples of this belief can be found in the works of Homer (circa 800 BCE).

Source of Rain

In the ancient near east, Hebrew scholars observed that even though the rivers ran into the sea, the sea never became full (Ecclesiastes 1:7). Some scholars conclude that the water cycle was described completely during this time in this passage: "The wind goeth toward the south, and turneth about unto the north; it whirleth about continually, and the wind returneth again according to its circuits. All the rivers run into the sea; yet the sea is not full; unto the place from whence the rivers come, thither they return again" (Ecclesiastes 1:6-7, KJV). Scholars are not in agreement as to the date of Ecclesiastes, though most scholars point to a date during the time of Solomon, the son of David and Bathsheba, "three thousand years ago, there is some agreement that the time period is 962-922 BCE. Furthermore, it was also observed that when the clouds were full, they emptied rain on the earth (Ecclesiastes 11:3). In addition, during 793-740 BC a Hebrew prophet, Amos, stated that water comes from the sea and is poured out on the earth (Amos 5:8, 9:6).

Precipitation and Percolation

In the Adityahridayam (a devotional hymn to the Sun God) of Ramayana, a Hindu epic dated to the 4th century BC, it is mentioned in the 22nd verse that the Sun heats up water and sends it down as rain. By roughly 500 BCE, Greek scholars were speculating that much of the water in rivers can be attributed to rain. The origin of rain was also known by then. These scholars maintained the belief, however, that water rising up through the earth contributed a great deal to rivers. Examples of this thinking included Anaximander (570 BCE) (who also speculated about the evolution of land animals from fish) and Xenophanes of Colophon (530 BCE). Chinese scholars such as Chi Ni Tzu (320 BC) and Lu Shih Ch'un Ch'iu (239 BCE) had similar thoughts. The idea that the water cycle is a closed cycle can be found in the works of Anaxagoras of Clazomenae (460 BCE) and Diogenes of Apollonia (460 BCE). Both Plato (390 BCE) and Aristotle (350 BCE) speculated about percolation as part of the water cycle.

Precipitation Alone

In the Biblical Book of Job, dated between 7th and 2nd centuries BCE, there is a description of precipitation in the hydrologic cycle, "For he maketh small the drops of water: they pour down rain according to the vapour thereof; Which the clouds do drop and distil upon man abundantly" (Job 36:27-28, KJV). Also found in the book of Ecclesiastes "All the rivers flow into the sea, Yet the sea is not full. To the place where the rivers flow, There they flow again." (Ecclesiastes 1:7)

Up to the time of the Renaissance, it was thought that precipitation alone was insufficient to feed rivers, for a complete water cycle, and that underground water pushing upwards from the oceans were the main contributors to river water. Bartholomew of England held this view (1240 CE), as did Leonardo da Vinci (1500 CE) and Athanasius Kircher (1644 CE).

The first published thinker to assert that rainfall alone was sufficient for the maintenance of rivers was Bernard Palissy (1580 CE), who is often credited as the "discoverer" of the modern theory of the water cycle. Palissy's theories were not tested scientifically until 1674, in a study commonly attributed to Pierre Perrault. Even then, these beliefs were not accepted in mainstream science until the early nineteenth century.

Water Distribution on Earth

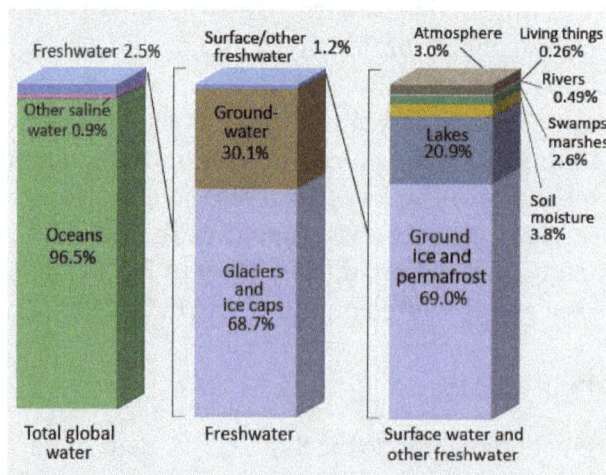

A graphical distribution of the locations of water on Earth

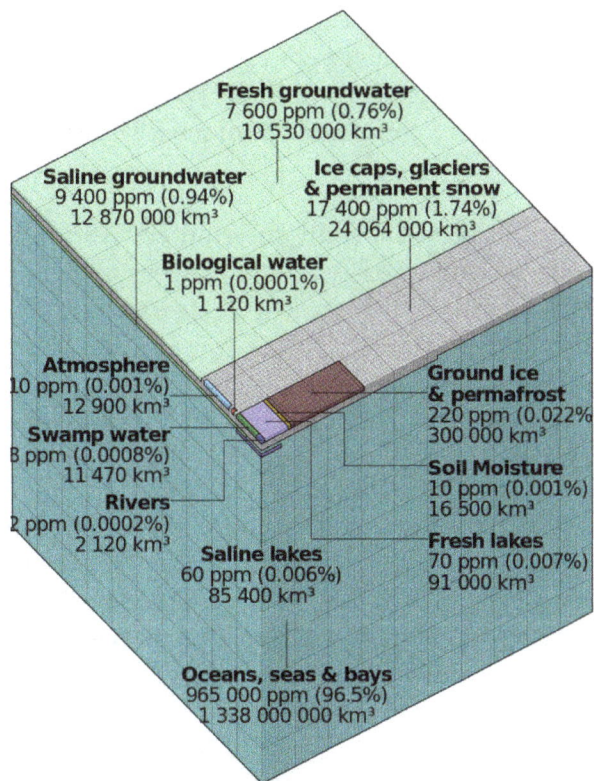

Visualisation of the distribution (by volume) of water on Earth. Each tiny cube (such as the one representing biological water) corresponds to approximately 1,000 km³ of water, with a mass of about 1 trillion tonnes (200,000 times that of the Great Pyramid of Giza or five times that of Lake Kariba, arguably the heaviest man-made object). Comprising 1 million tiny cubes, the entire cube would measure about 1,102 km on each side.

The water distribution on earth shows that most water in the Earth's atmosphere and crust comes from the world ocean's saline seawater, while freshwater accounts for only 2.5% of the total. Because the oceans that cover roughly 71% of the area of the Earth reflect blue light, the Earth appears blue from space, and is often referred to as the *blue planet* and the *Pale Blue Dot*. An estimated 1.5 to 11 times the amount of water in the oceans may be found hundreds of miles deep within the Earth's interior, although not in liquid form.

The oceanic crust is young, thin and dense, with none of the rocks within it dating from any older than the breakup of Pangaea. Because water is much denser than any gas, this means that water will flow into the "depressions" formed as a result of the high density of oceanic crust. (On a planet like Venus, with no water, the depressions appear to form a vast plain above which rise plateaux). Since the low density rocks of the continental crust contain large quantities of easily eroded salts of the alkali and alkaline earth metals, salt has, over billions of years, accumulated in the oceans as a result of evaporation returning the fresh water to land as rain and snow.

As a result, the vast bulk of the water on Earth is regarded as *saline* or *salt water*, with an average salinity of 35‰ (or 3.5%, roughly equivalent to 34 grams of salts in 1 kg of seawater), though this varies slightly according to the amount of runoff received from surrounding land. In all, water from oceans and marginal seas, saline groundwater and water from saline closed lakes amount to over 97% of the water on Earth, though no closed lake stores a globally significant amount of water. *Saline* groundwater is seldom considered except when evaluating water quality in arid regions.

The remainder of the Earth's water constitutes the planet's *fresh water* resource. Typically, fresh water is defined as water with a salinity of *less than 1 percent that of the oceans* - i.e. below around 0.35‰. Water with a salinity between this level and 1‰ is typically referred to as *marginal water* because it is marginal for many uses by humans and animals. The ratio of salt water to fresh water on Earth is around 40 to 1.

The planet's fresh water is also very unevenly distributed. Although in warm periods such as the Mesozoic and Paleogene when there were no glaciers anywhere on the planet all fresh water was found in rivers and streams, today most fresh water exists in the form of ice, snow, groundwater and soil moisture, with only 0.3% in liquid form on the surface. Of the liquid surface fresh water, 87% is contained in lakes, 11% in swamps, and only 2% in rivers. Small quantities of water also exist in the atmosphere and in living beings. Of these sources, only river water is generally valuable.

Most lakes are in very inhospitable regions such as the glacial lakes of Canada, Lake Baikal in Russia, Lake Khövsgöl in Mongolia, and the African Great Lakes. The North American Great Lakes, which contain 21% of the world's fresh water by volume, are the exception. They are located in a hospitable region, which is heavily populated. The Great Lakes Basin is home to 33 million people. The Canadian cities of Toronto, Hamilton, Ontario, St. Catharines, Niagara, Oshawa, Windsor, and Barrie, and the United States cities of Duluth, Milwaukee, Chicago, Gary, Detroit, Cleveland, Buffalo, and Rochester, are all located on shores of the Great Lakes.

Although the total volume of groundwater is known to be much greater than that of river runoff, a large proportion of this groundwater is saline and should therefore be classified with the saline water above. There is also a lot of *fossil* groundwater in arid regions that has never been renewed for thousands of years; this must not be seen as renewable water.

However, fresh groundwater is of great value, especially in arid countries such as India. Its distribution is broadly similar to that of surface river water, but it is easier to store in hot and dry climates because groundwater storages are much more shielded from evaporation than are dams. In countries such as Yemen, groundwater from erratic rainfall during the rainy season is the major source of irrigation water.

Because groundwater recharge is much more difficult to accurately measure than surface runoff, groundwater is not generally used in areas where even fairly limited levels of surface water are available. Even today, estimates of total groundwater recharge vary greatly for the same region depending on what source is used, and cases where fossil groundwater is exploited beyond the recharge rate (including the Ogallala Aquifer) are very frequent and almost always not seriously considered when they were first developed.

Distribution of Saline and Fresh Water

The total volume of water on Earth is estimated at 1.386 billion km³ (333 million cubic miles), with 97.5% being salt water and 2.5% being fresh water. Of the fresh water, only 0.3% is in liquid form on the surface. In addition, the lower mantle of inner earth may hold as much as 5 times more water than all surface water combined (all oceans, all lakes, all rivers).

Source of Water	Volume of Water in km³ (cu mi)	% Total Water	% Salt Water	% Fresh Water	% Liquid Surface Fresh Water
Oceans	1,338,000,000 (321,000,000)	96.5	99.0		
Pacific Ocean	669,880,000 (160,710,000)	48.3	49.6		
Atlantic Ocean	310,410,900 (74,471,500)	22.4	23.0		
Indian Ocean	264,000,000 (63,000,000)	19.0	19.5		
Southern Ocean	71,800,000 (17,200,000)	5.18	5.31		
Arctic Ocean	18,750,000 (4,500,000)	1.35	1.39		
Ice and snow	24,364,000 (5,845,000)	1.76		69.6	
Glaciers	24,064,000 (5,773,000)	1.74		68.7	
Antarctic ice sheet	21,600,000 (5,200,000)	1.56		61.7	
Greenland ice sheet	2,340,000 (560,000)	0.17		6.68	
Arctic islands	83,500 (20,000)	0.006		0.24	
Mountain ranges	40,600 (9,700)	0.003		0.12	
Ground ice and permafrost	300,000 (72,000)	0.022		0.86	
Groundwater	23,400,000 (5,600,000)	1.69			
Saline groundwater	12,870,000 (3,090,000)	0.93	0.95		
Fresh groundwater	10,530,000 (2,530,000)	0.76		30.1	
Soil moisture	16,500 (4,000)	0.0012		0.047	
Lakes	176,400 (42,300)	0.013			
Saline lakes	85,400 (20,500)	0.0062	0.0063		
Caspian Sea	78,200 (18,800)	0.0056	0.0058		
Other saline lakes	7,200 (1,700)	0.00052	0.00053		
Fresh water lakes	91,000 (22,000)	0.0066		0.26	87.0
African Great Lakes	30,070 (7,210)	0.0022		0.086	28.8
Lake Baikal	23,615 (5,666)	0.0017		0.067	22.6
North American Great Lakes	22,115 (5,306)	0.0016		0.063	21.1
Other fresh water lakes	15,200 (3,600)	0.0011		0.043	14.5
Swamps	11,470 (2,750)	0.00083		0.033	11.0
Rivers	2,120 (510)	0.00015		0.0061	2.03
Biological water	1,120 (270)	0.000081		0.0032	
Atmosphere	12,900 (3,100)	0.00093		0.037	

Distribution of River Water

The total volume of water in rivers is estimated at 2,120 km³ (510 cubic miles), or 2% of the surface fresh water on Earth. Rivers and basins are often compared not according to their static volume, but to their flow of water, or surface runoff. The distribution of river runoff across the Earth's surface is very uneven.

Continent or region	River runoff (km³/year)	Percent of world total
North America	7,800	17.9
South America	12,000	27.6
Europe	2,900	6.7
Middle East and North Africa	140	0.3
Sub-Saharan Africa	4,000	9.2
Asia (excluding Middle East)	13,300	30.6
Australia	440	1.0
Oceania	6,500	14.9

There can be huge variations within these regions. For example, as much as a quarter of Australia's limited renewable fresh water supply is found in almost uninhabited Cape York Peninsula. Also, even in well-watered continents, there are areas that are extremely short of water, such as Texas in North America, whose renewable water supply totals only 26 km³/year in an area of 695,622 km², or South Africa, with only 44 km³/year in 1,221,037 km². The areas of greatest concentration of renewable water are:

- The Amazon and Orinoco Basins (a total of 6,500 km³/year or 15 percent of global runoff)

- East Asia

  o Yangtze Basin - 1,000 km³/year

- South and Southeast Asia, with a total of 8,000 km³/year or 18 percent of global runoff

  o Brahmaputra Basin - 900 km³/year

  o Irrawaddy Basin - 500 km³/year

  o Mekong Basin - 450 km³/year

- Canada, with over 10 percent of world's river water and large numbers in lakes

  o Mackenzie River - over 250 km³/year

  o Yukon River - over 150 km³/year

- Siberia

  o Yenisey - over 5% of world's fresh water in basin - second largest after the Amazon

  o Ob River - over 500 km³/year

  o Lena River - over 450 km³/year

- New Guinea

  o Fly and Sepik Rivers - total over 300 km³/year in only about 150,000 km² of basin area.

Area, Volume, and Depth of the World Ocean

Body of Water	Area (10^6 km^2)	Volume (10^6 km^3)	Mean Depth (m)
Atlantic Ocean	82.4	323.6	3,926
Pacific Ocean	165.2	707.6	4,282
Indian Ocean	73.4	291.0	3,963
All oceans and seas	361	1,370	3,796

Variability of Water Availability

Variability of water availability is important both for the functioning of aquatic species and also for the availability of water for human use: water that is only available in a few wet years must not be considered renewable. Because most global runoff comes from areas of very low climatic variability, the total global runoff is generally of low variability.

Indeed, even in most arid zones, there tends to be few problems with variability of runoff because most usable sources of water come from high mountain regions which provide highly reliable glacier melt as the chief source of water, which also comes in the summer peak period of high demand for water. This historically aided the development of many of the great civilizations of ancient history, and even today allows for agriculture in such productive areas as the San Joaquin Valley.

However, in Australia and Southern Africa, the story is different. Here, runoff variability is much higher than in other continental regions of the world with similar climates. Typically temperate (Köppen climate classification C) and arid (Köppen climate classification B) climate rivers in Australia and Southern Africa have as much as three times the coefficient of variation of runoff of those in other continental regions. The reason for this is that, whereas all other continents have had their soils largely shaped by Quaternary glaciation and mountain building, soils of Australia and Southern Africa have been largely unaltered since at least the early Cretaceous and generally since the previous ice age in the Carboniferous. Consequently, available nutrient levels in Australian and Southern African soils tend to be orders of magnitude lower than those of similar climates in other continents, and native flora compensate for this through much higher rooting densities (e.g. proteoid roots) to absorb minimal phosphorus and other nutrients. Because these roots absorb so much water, runoff in typical Australian and Southern African rivers does not occur until about 300 mm (12 inches) or more of rainfall has occurred. In other continents, runoff will occur after quite light rainfall due to the low rooting densities.

Climate type (Köppen)	Mean annual rainfall	Typical runoff ratio for Australia and Southern Africa	Typical runoff ratio for rest of the world
BWh	250 mm (10 inches)	1 percent (2.5 mm)	10 percent (25 mm)
BSh (on Mediterranean fringe)	350 mm (14 inches)	3 percent (12 mm)	20 percent (80 mm)
Csa	500 mm (20 inches)	5 percent (25 mm)	35 percent (175 mm)
Caf	900 mm (36 inches)	15 percent (150 mm)	45 percent (400 mm)
Cb	1100 mm (43 inches)	25 percent (275 mm)	70 percent (770 mm)

The consequence of this is that many rivers in Australia and Southern Africa (as compared to *extremely few* in other continents) are theoretically impossible to regulate because rates of evaporation from dams mean a storage sufficiently large to theoretically regulate the river to a given level would actually allow very little draft to be used. Examples of such rivers include those in the Lake Eyre Basin. Even for other Australian rivers, a storage three times as large is needed to provide a third the supply of a comparable climate in southeastern North America or southern China. It also affects aquatic life, favouring strongly those species able to reproduce rapidly after high floods so that some will survive the next drought.

Tropical (Köppen climate classification A) climate rivers in Australia and Southern Africa do not, in contrast, have markedly lower runoff ratios than those of similar climates in other regions of the world. Although soils in tropical Australia and southern Africa are even poorer than those of the arid and temperate parts of these continents, vegetation can use organic phosphorus or phosphate dissolved in rainwater as a source of the nutrient. In cooler and drier climates these two related sources tend to be virtually useless, which is why such specialised means are needed to extract the most minimal phosphorus.

There are other isolated areas of high runoff variability, though these are basically due to erratic rainfall rather than different hydrology. These include:

- Southwest Asia

- The Brazilian Nordeste

- The Great Plains of the United States

Water in Earth's Mantle

It is estimated an additional 1.5 to eleven times the amount of water in the oceans is contained in the Earth's interior, and some scientists have hypothesized that the water in the mantle is part of a "whole-Earth water cycle". Some of the water in the mantle is dissolved in various minerals near the transition zone between Earth's upper and lower mantle. At temperatures of 1,100 °C (2,010 °F) and extreme pressures found deep underground, water breaks down into hydroxyls and oxygen. The existence of water was experimentally predicted in 2002, and direct evidence of the water was found in 2014 based on tests on a sample of ringwoodite. Further evidence for large quantities of water in the mantle was found in observations of melting in the transition zone from the USArray project. Liquid water is not present within the ringwoodite, rather the components of water (hydrogen and oxygen) are held within as hydroxide ions.

Water Quality

Water quality refers to the chemical, physical, biological, and radiological characteristics of water. It is a measure of the condition of water relative to the requirements of one or more biotic species and or to any human need or purpose. It is most frequently used by reference to a set of standards against which compliance can be assessed. The most common standards used to assess water quality relate to health of ecosystems, safety of human contact, and drinking water.

A rosette sampler is used to collect water samples in deep water, such as the Great Lakes or oceans, for water quality testing.

Standards

In the setting of standards, agencies make political and technical/scientific decisions about how the water will be used. In the case of natural water bodies, they also make some reasonable estimate of pristine conditions. Different uses raise different concerns and therefore different standards are considered. Natural water bodies will vary in response to environmental conditions. Environmental scientists work to understand how these systems function, which in turn helps to identify the sources and fates of contaminants. Environmental lawyers and policymakers work to define legislation with the intention that water is maintained at an appropriate quality for its identified use.

The vast majority of surface water on the planet is neither potable nor toxic. This remains true when seawater in the oceans (which is too salty to drink) is not counted. Another general perception of *water quality* is that of a simple property that tells whether water is polluted or not. In fact, water quality is a complex subject, in part because water is a complex medium intrinsically tied to the ecology of the Earth. Industrial and commercial activities (e.g. manufacturing, mining, construction, transport) are a major cause of water pollution as are runoff from agricultural areas, urban runoff and discharge of treated and untreated sewage.

Categories

The parameters for water quality are determined by the intended use. Work in the area of water quality tends to be focused on water that is treated for human consumption, industrial use, or in the environment.

Human Consumption

Contaminants that may be in untreated water include microorganisms such as viruses, protozoa and bacteria; inorganic contaminants such as salts and metals; organic chemical contaminants from industrial processes and petroleum use; pesticides and herbicides; and radioactive contaminants. Water quality depends on the local geology and ecosystem, as well as human uses such as sewage dispersion, industrial pollution, use of water bodies as a heat sink, and overuse (which may lower the level of the water).

The United States Environmental Protection Agency (EPA) limits the amounts of certain contaminants in tap water provided by US public water systems. The Safe Drinking Water Act authorizes EPA to issue two types of standards: *primary standards* regulate substances that potentially affect human health, and *secondary standards* prescribe aesthetic qualities, those that affect taste, odor, or appearance. The U.S. Food and Drug Administration (FDA) regulations establish limits for contaminants in bottled water that must provide the same protection for public health. Drinking water, including bottled water, may reasonably be expected to contain at least small amounts of some contaminants. The presence of these contaminants does not necessarily indicate that the water poses a health risk.

In urbanized areas around the world, water purification technology is used in municipal water systems to remove contaminants from the source water (surface water or groundwater) before it is distributed to homes, businesses, schools and other recipients. Water drawn directly from a stream, lake, or aquifer and that has no treatment will be of uncertain quality.

Industrial and Domestic Use

Dissolved minerals may affect suitability of water for a range of industrial and domestic purposes. The most familiar of these is probably the presence of ions of calcium and magnesium which interfere with the cleaning action of soap, and can form hard sulfate and soft carbonate deposits in water heaters or boilers. Hard water may be softened to remove these ions. The softening process often substitutes sodium cations. Hard water may be preferable to soft water for human consumption, since health problems have been associated with excess sodium and with calcium and magnesium deficiencies. Softening decreases nutrition and may increase cleaning effectiveness.

Environmental Water Quality

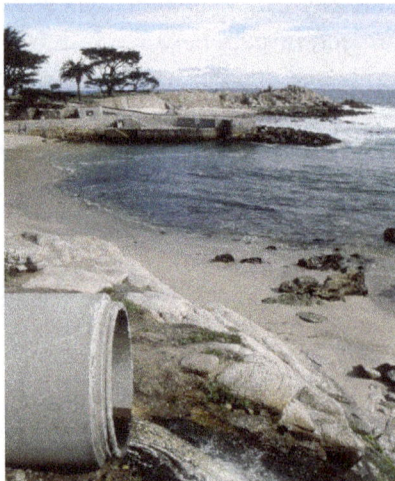

Urban runoff discharging to coastal waters

Environmental water quality, also called ambient water quality, relates to water bodies such as lakes, rivers, and oceans. Water quality standards for surface waters vary significantly due to different environmental conditions, ecosystems, and intended human uses. Toxic substances and high populations of certain microorganisms can present a health hazard for non-drinking purposes such as irrigation, swimming, fishing, rafting, boating, and industrial uses. These conditions may also affect wildlife, which use the water for drinking or as a habitat. Modern water quality laws generally specify protection of fisheries and recreational use and require, as a minimum, retention of current quality standards.

Satirical cartoon by William Heath, showing a woman observing monsters in a drop of London water
(at the time of the *Commission on the London Water Supply* report, 1828)

There is some desire among the public to return water bodies to pristine, or pre-industrial conditions. Most current environmental laws focus on the designation of particular uses of a water body. In some countries these designations allow for some water contamination as long as the particular type of contamination is not harmful to the designated uses. Given the landscape changes (e.g., land development, urbanization, clearcutting in forested areas) in the watersheds of many freshwater bodies, returning to pristine conditions would be a significant challenge. In these cases, environmental scientists focus on achieving goals for maintaining healthy ecosystems and may concentrate on the protection of populations of endangered species and protecting human health.

Sampling and Measurement

The complexity of water quality as a subject is reflected in the many types of measurements of water quality indicators. The most accurate measurements of water quality are made on-site, because water exists in equilibrium with its surroundings. Measurements commonly made on-site and in direct contact with the water source in question include temperature, pH, dissolved oxygen, conductivity, oxygen reduction potential (ORP), turbidity, and Secchi disk depth.

Sample Collection

More complex measurements are often made in a laboratory requiring a water sample to be collected, preserved, transported, and analyzed at another location. The process of water sampling introduces two significant problems. The first problem is the extent to which the sample may be representative of the water source of interest. Many water sources vary with time and with location. The measurement of interest may vary seasonally or from day to night or in response to some activity of man or natural populations of aquatic plants and animals. The measurement of interest may vary with distances from the water boundary with overlying atmosphere and underlying or confining soil. The sampler must determine if a single time and location meets the needs of the investigation, or if the water use of interest can be satisfactorily assessed by averaged values with time and/or location, or if critical maxima and minima require individual measurements over a range of times, locations and/or events. The sample collection procedure must assure correct weighting of individual sampling times and locations where averaging is appropriate.[39-40] Where critical maximum or minimum values exist, statistical methods must be applied to observed variation to determine an adequate number of samples to assess probability of exceeding those critical values.

An automated sampling station installed along the East Branch Milwaukee River, New Fane, Wisconsin. The cover of the 24-bottle autosampler (center) is partially raised, showing the sample bottles inside. The autosampler was programmed to collect samples at time intervals, or proportionate to flow over a specified period. The data logger (white cabinet) recorded temperature, specific conductance, and dissolved oxygen levels.

The second problem occurs as the sample is removed from the water source and begins to establish chemical equilibrium with its new surroundings - the sample container. Sample containers must be made of materials with minimal reactivity with substances to be measured; and pre-cleaning of sample containers is important. The water sample may dissolve part of the sample container and any residue on that container, or chemicals dissolved in the water sample may sorb onto the sample container and remain there when the water is poured out for analysis.[4] Similar physical and chemical interactions may take place with any pumps, piping, or intermediate devices used to transfer the water sample into the sample container. Water collected from depths below the surface will normally be held at the reduced pressure of the atmosphere; so gas dissolved in the water may escape into unfilled space at the top of the container. Atmospheric gas present in that air space may also dissolve into the water sample. Other chemical reaction equilibria may change if the water sample changes temperature. Finely divided solid particles formerly suspended by water turbulence may settle to the bottom of the sample container, or a solid phase may form from biological growth or chemical precipitation. Microorganisms within the water sample may biochemically alter concentrations of oxygen, carbon dioxide, and organic compounds. Changing carbon dioxide concentrations may alter pH and change solubility of chemicals of interest. These problems are of special concern during measurement of chemicals assumed to be significant at very low concentrations.

Filtering a manually collected water sample (grab sample) for analysis

Sample preservation may partially resolve the second problem. A common procedure is keeping samples cold to slow the rate of chemical reactions and phase change, and analyzing the sample as soon as possible; but this merely minimizes the changes rather than preventing them.[43-45] A useful procedure for determining influence of sample containers during delay between sample collection and analysis involves preparation for two artificial samples in advance of the sampling event. One sample container is filled with water known from previous analysis to contain no detectable amount of the chemical of interest. This sample, called a "blank," is opened for exposure to the atmosphere when the sample of interest is collected, then resealed and transported to the laboratory with the sample for analysis to determine if sample holding procedures introduced any measurable amount of the chemical of interest. The second artificial sample is collected with the sample of interest, but then "spiked" with a measured additional amount of the chemical of interest at the time of collection. The blank and spiked samples are carried with the sample of interest and analyzed by the same methods at the same times to determine any changes indicating gains or losses during the elapsed time between collection and analysis.

Testing in Response to Natural Disasters and Other Emergencies

Inevitably after events such as earthquakes and tsunamis, there is an immediate response by the aid agencies as relief operations get underway to try and restore basic infrastructure and provide the basic fundamental items that are necessary for survival and subsequent recovery. Access to clean drinking water and adequate sanitation is a priority at times like this. The threat of disease increases hugely due to the large numbers of people living close together, often in squalid conditions, and without proper sanitation.

After a natural disaster, as far as water quality testing is concerned there are widespread views on the best course of action to take and a variety of methods can be employed. The key basic water quality parameters that need to be addressed in an emergency are bacteriological indicators of fecal contamination, free chlorine residual, pH, turbidity and possibly conductivity/total dissolved solids. There are a number of portable water test kits on the market widely used by aid and relief agencies for carrying out such testing.

After major natural disasters, a considerable length of time might pass before water quality returns to pre-disaster levels. For example, following the 2004 Indian Ocean Tsunami the Colombo-based International Water Management Institute (IWMI) monitored the effects of saltwater and concluded that the wells recovered to pre-tsunami drinking water quality one and a half years after the event. IWMI developed protocols for cleaning wells contaminated by saltwater; these were subsequently officially endorsed by the World Health Organization as part of its series of Emergency Guidelines.

Chemical Analysis

A gas chromatograph-mass spectrometer measures pesticides and other organic pollutants

The simplest methods of chemical analysis are those measuring chemical elements without respect to their form. Elemental analysis for oxygen, as an example, would indicate a concentration of 890,000 milligrams per litre (mg/L) of water sample because water is made of oxygen. The method selected to measure dissolved oxygen should differentiate between diatomic oxygen and oxygen combined with other elements. The comparative simplicity of elemental analysis has produced a large amount of sample data and water quality criteria for elements sometimes identified as heavy metals. Water analysis for heavy metals must consider soil particles suspended in the water sample. These suspended soil particles may contain measurable amounts of metal. Although the particles are not dissolved in the water, they may be consumed by people drinking the water. Adding acid to a water sample to prevent loss of dissolved metals onto the sample container may dissolve more metals from suspended soil particles. Filtration of soil particles from the water sample before acid addition, however, may cause loss of dissolved metals onto the filter. The complexities of differentiating similar organic molecules are even more challenging.

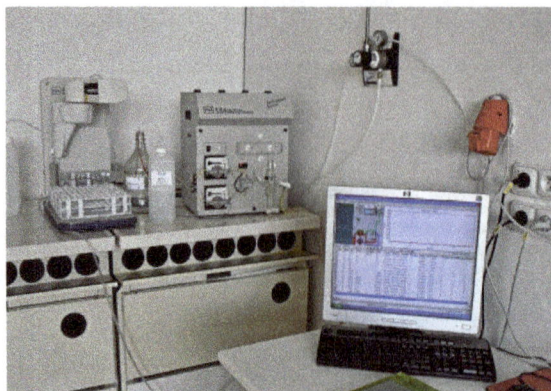

Atomic fluorescence spectroscopy is used to measure mercury and other heavy metals

Making these complex measurements can be expensive. Because direct measurements of water quality can be expensive, ongoing monitoring programs are typically conducted by government agencies. However, there are local volunteer programs and resources available for some general assessment. Tools available to the general public include on-site test kits, commonly used for home fish tanks, and biological assessment procedures.

Real-time Monitoring

Although water quality is usually sampled and analyzed at laboratories, nowadays, citizens demand real-time information about the water they are drinking. During the last years, several companies are deploying worldwide real-time remote monitoring systems for measuring water pH, turbidity or dissolved oxygen levels.

Drinking Water Indicators

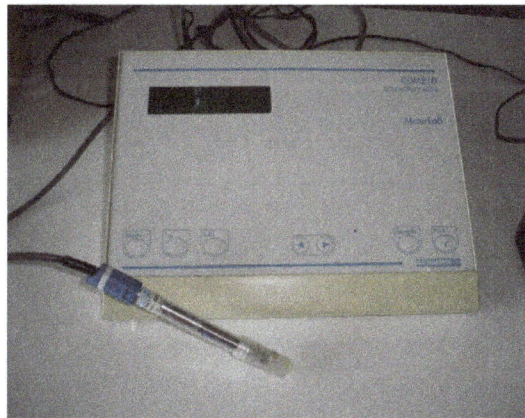

An electrical conductivity meter is used to measure total dissolved solids

The following is a list of indicators often measured by situational category:

- Alkalinity

- Color of water

- pH

- Taste and odor (geosmin, 2-Methylisoborneol (MIB), etc.)

- Dissolved metals and salts (sodium, chloride, potassium, calcium, manganese, magnesium)

- Microorganisms such as fecal coliform bacteria (*Escherichia coli*), Cryptosporidium, and Giardia lamblia; *see* Bacteriological water analysis

- Dissolved metals and metalloids (lead, mercury, arsenic, etc.)

- Dissolved organics: colored dissolved organic matter (CDOM), dissolved organic carbon (DOC)

- Radon

- Heavy metals

- Pharmaceuticals
- Hormone analogs

Environmental Indicators

Physical Indicators

• Water Temperature • Specifics Conductance or EC, Electrical Conductance, Conductivity • Total suspended solids (TSS) • Transparency or Turbidity	• Total dissolved solids (TDS) • Odour of water • Color of water • Taste of water

Chemical Indicators

• pH • Biochemical oxygen demand (BOD) • Chemical oxygen demand (COD) • Dissolved oxygen (DO) • Total hardness (TH)	• Heavy metals • Nitrate • Orthophosphates • Pesticides • Surfactants

Biological Indicators

• Ephemeroptera • Plecoptera • Mollusca • Trichoptera	• *Escherichia coli* (E. coli) • Coliform bacteria

Biological monitoring metrics have been developed in many places, and one widely used measure is the presence and abundance of members of the insect orders Ephemeroptera, Plecoptera and Trichoptera. (Common names are, respectively, Mayfly, Stonefly and Caddisfly.) EPT indexes will naturally vary from region to region, but generally, within a region, the greater the number of taxa from these orders, the better the water quality. Organisations in the United States, such as EPA offer guidance on developing a monitoring program and identifying members of these and other aquatic insect orders.

Individuals interested in monitoring water quality who cannot afford or manage lab scale analysis can also use biological indicators to get a general reading of water quality. One example is the IOWATER volunteer water monitoring program, which includes a benthic macroinvertebrate indicator key.

Bivalve molluscs are largely used as bioindicators to monitor the health of aquatic environments in both fresh water and the marine environments. Their population status or structure, physiology, behaviour or the level of contamination with elements or compounds can indicate the state of contamination status of the ecosystem. They are particularly useful since they are sessile so that they are representative of the environment where they are sampled or placed. A typical project is the Mussel Watch Programme, but today they are used worldwide.

The Southern African Scoring System (SASS) method is a biological water quality monitoring system based on the presence of benthic macroinvertebrates. The SASS aquatic biomonitoring tool has been refined over the past 30 years and is now on the fifth version (SASS5) which has been specifically modified in accordance with international standards, namely the ISO/IEC 17025 protocol. The SASS5 method is used by the South African Department of Water Affairs as a standard method for River Health Assessment, which feeds the national River Health Programme and the national Rivers Database.

Standards and Reports

International

- The World Health Organisation (WHO) has published guidelines for drinking-water quality (GDWQ) in 2011.

- The International Organization for Standardization (ISO) published regulation of water quality in the section of ICS 13.060, ranging from water sampling, drinking water, industrial class water, sewage, and examination of water for chemical, physical or biological properties. ICS 91.140.60 covers the standards of water supply systems.

National specifications for ambient water and drinking water

European Union

The water policy of the European Union is primarily codified in three directives:

- Directive on Urban Waste Water Treatment (91/271/EEC) of 21 May 1991 concerning discharges of municipal and some industrial wastewaters;

- The Drinking Water Directive (98/83/EC) of 3 November 1998 concerning potable water quality;

- Water Framework Directive (2000/60/EC) of 23 October 2000 concerning water resources management.

India

- Indian Council of Medical Research (ICMR) Standards for Drinking Water.

South Africa

Water quality guidelines for South Africa are grouped according to potential user types (e.g. domestic, industrial) in the 1996 Water Quality Guidelines. Drinking water quality is subject to the South African National Standard (SANS) 241 Drinking Water Specification.

United Kingdom

In England and Wales acceptable levels for drinking water supply are listed in the "Water Supply (Water Quality) Regulations 2000."

United States

In the United States, Water Quality Standards are defined by state agencies for various water bodies, guided by the desired uses for the water body (e.g., fish habitat, drinking water supply, recreational use). The Clean Water Act (CWA) requires each governing jurisdiction (states, territories, and covered tribal entities) to submit a set of biennial reports on the quality of water in their area. These reports are known as the 303(d) and 305(b) reports, named for their respective CWA provisions, and are submitted to, and approved by, EPA. These reports are completed by the governing jurisdiction, typically a state environmental agency. In coming years it is expected that the governing jurisdictions will submit all three reports as a single document, called the "Integrated Report." The *National Water Quality Inventory Report to Congress* is a general report on water quality, providing overall information about the number of miles of streams and rivers and their aggregate condition. The CWA requires states to adopt standards for each of the possible designated uses that they assign to their waters. Should evidence suggest or document that a stream, river or lake has failed to meet the water quality criteria for one or more of its designated uses, it is placed on a list of impaired waters. Once a state has placed a water body on this list, it must develop a management plan establishing Total Maximum Daily Loads for the pollutant(s) impairing the use of the water. These TMDLs establish the reductions needed to fully support the designated uses.

Drinking water standards, which are applicable to public water systems, are issued by EPA under the Safe Drinking Water Act.

Fresh Water

Fresh water is naturally occurring water on Earth's surface in ice sheets, ice caps, glaciers, icebergs, bogs, ponds, lakes, rivers and streams, and underground as groundwater in aquifers and underground streams. Fresh water is generally characterized by having low concentrations of dissolved salts and other total dissolved solids. The term specifically excludes seawater and brackish water although it does include mineral-rich waters such as chalybeate springs. The term "sweet water" (from Spanish "agua dulce") has been used to describe fresh water in contrast to salt water. The term fresh water does not have the same meaning as potable water. Much of the surface fresh water and ground water is unsuitable for drinking without some form of purification because of the presence of chemical or biological contaminants.

Earth seen from Apollo 17 — the Antarctic ice sheet at the bottom of the photograph contains 61% of the fresh water, or 1.7% of the total water, on Earth.

Systems

Rivers, lakes, and marshlands, such as (from top) South America's Amazon River, Russia's Lake Baikal, and the Everglades in the United States, are types of freshwater systems.

Fresh water habitats are divided into lentic systems, which are the stillwaters including ponds, lakes, swamps and mires; lotic, or running-water systems; and groundwater which flows in rocks and aquifers. There is, in addition, a zone which bridges between groundwater and lotic systems, which is the hyporheic zone, which underlies many larger rivers and can contain substantially more water than is seen in the open channel. It may also be in direct contact with the underlying underground water. The majority of fresh water on Earth is in ice caps.

Sources

The source of almost all fresh water is precipitation from the atmosphere, in the form of mist, rain and snow. Fresh water falling as mist, rain or snow contains materials dissolved from the atmosphere and material from the sea and land over which the rain bearing clouds have traveled. In industrialized areas rain is typically acidic because of dissolved oxides of sulfur and nitrogen formed from burning of fossil fuels in cars, factories, trains and aircraft and from the atmospheric emissions of industry. In some cases this acid rain results in pollution of lakes and rivers.

In coastal areas fresh water may contain significant concentrations of salts derived from the sea if windy conditions have lifted drops of seawater into the rain-bearing clouds. This can give rise to elevated concentrations of sodium, chloride, magnesium and sulfate as well as many other compounds in smaller concentrations.

In desert areas, or areas with impoverished or dusty soils, rain-bearing winds can pick up sand and dust and this can be deposited elsewhere in precipitation and causing the freshwater flow to be measurably contaminated both by insoluble solids but also by the soluble components of those soils. Significant quantities of iron may be transported in this way including the well-documented transfer of iron-rich rainfall falling in Brazil derived from sand-storms in the Sahara in north Africa.

Water Distribution

Water is a critical issue for the survival of all living organisms. Some can use salt water but many organisms including the great majority of higher plants and most mammals must have access to fresh water to live. Some terrestrial mammals, especially desert rodents appear to survive without drinking but they do generate water through the metabolism of cereal seeds and they also have mechanisms to conserve water to the maximum degree.

Out of all the water on Earth, saline water in oceans, seas and saline groundwater make up about 97% of it. Only 2.5–2.75% is fresh water, including 1.75–2% frozen in glaciers, ice and snow, 0.5–0.75% as fresh groundwater and soil moisture, and less than 0.01% of it as surface water in lakes, swamps and rivers. Freshwater lakes contain about 87% of this fresh surface water, including 29% in the African Great Lakes, 20% in Lake Baikal in Russia, 21% in the North American Great Lakes, and 14% in other lakes. Swamps have most of the balance with only a small amount in rivers, most notably the Amazon River. The atmosphere contains 0.04% water. In areas with no fresh water on the ground surface, fresh water derived from precipitation may, because of its lower density, overlie saline ground water in lenses or layers. Most of the world's fresh water is frozen in ice sheets. Many areas suffer from lack of distribution of fresh water, such as deserts.

Fresh groundwater
7 600 ppm (0.76%)
10 530 000 km³

Saline groundwater
9 400 ppm (0.94%)
12 870 000 km³

Ice caps, glaciers & permanent snow
17 400 ppm (1.74%)
24 064 000 km³

Biological water
1 ppm (0.0001%)
1 120 km³

Atmosphere
10 ppm (0.001%)
12 900 km³

Ground ice & permafrost
220 ppm (0.022%)
300 000 km³

Swamp water
8 ppm (0.0008%)
11 470 km³

Soil Moisture
10 ppm (0.001%)
16 500 km³

Rivers
2 ppm (0.0002%)
2 120 km³

Fresh lakes
70 ppm (0.007%)
91 000 km³

Saline lakes
60 ppm (0.006%)
85 400 km³

Oceans, seas & bays
965 000 ppm (96.5%)
1 338 000 000 km³

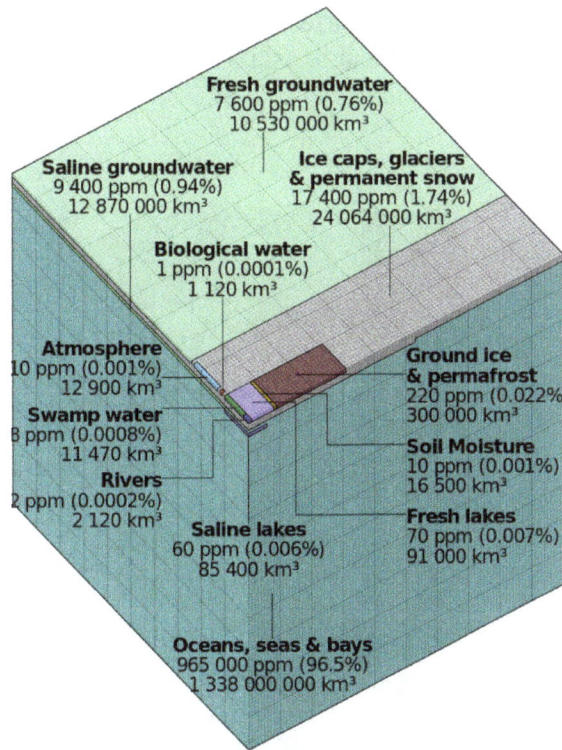

Visualisation of the distribution (by volume) of water on Earth. Each tiny cube (such as the one representing biological water) corresponds to approximately 1000 cubic km of water, with a mass of approximately 1 trillion tonnes (200000 times that of the Great Pyramid of Giza or 5 times that of Lake Kariba, arguably the heaviest man-made object). The entire block comprises 1 million tiny cubes.

Numerical Definition

Fresh water can be defined as water with less than 500 parts per million (ppm) of dissolved salts.

Water salinity based on dissolved salts			
Fresh water	Brackish water	Saline water	Brine
< 0.05%	0.05–3%	3–5%	> 5%

Other sources give higher upper salinity limits for fresh water, e.g. 1000 ppm or 3000 ppm.

Aquatic Organisms

Fresh water creates a hypotonic environment for aquatic organisms. This is problematic for some organisms with pervious skins or with gill membranes, whose cell membranes may burst if excess water is not excreted. Some protists accomplish this using contractile vacuoles, while freshwater fish excrete excess water via the kidney. Although most aquatic organisms have a limited ability to regulate their osmotic balance and therefore can only live within a narrow range of salinity, diadromous fish have the ability to migrate between fresh water and saline water bodies. During these migrations they undergo changes to adapt to the surroundings of the changed salinities; these processes are hormonally controlled. The eel (*Anguilla anguilla*) uses the hormone prolactin, while in salmon (*Salmo salar*) the hormone cortisol plays a key role during this process.

Many sea birds have special glands at the base of the bill through which excess salt is excreted. Similarly the marine iguanas on the Galápagos Islands excrete excess salt through a nasal gland and they sneeze out a very salty excretion.

Fresh Water as a Resource

Water fountain found in a small Swiss village. They are used as a drinking water source for people and cattle. Almost every Alpine village has such a water source.

An important concern for hydrological ecosystems is securing minimum streamflow, especially preserving and restoring instream water allocations. Fresh water is an important natural resource necessary for the survival of all ecosystems. The use of water by humans for activities such as irrigation and industrial applications can have adverse impacts on down-stream ecosystems. Chemical contamination of fresh water can also seriously damage eco-systems.

Pollution from human activity, including oil spills and also presents a problem for freshwater resources. The largest petroleum spill that has ever occurred in fresh water was caused by a Royal Dutch Shell tank ship in Magdalena, Argentina, on 15 January 1999, polluting the environment, drinkable water, plants and animals.

Fresh and unpolluted water accounts for 0.003% of total water available globally.

Agriculture

Changing landscape for the use of agriculture has a great effect on the flow of fresh water. Changes in landscape by the removal of trees and soils changes the flow of fresh water in the local environment and also affects the cycle of fresh water. As a result, more fresh water is stored in the soil which benefits agriculture. However, since agriculture is the human activity that consumes the most fresh water, this can put a severe strain on local freshwater resources resulting in the destruction of local ecosystems. In Australia, over-abstraction of fresh water for intensive irrigation activities has caused 33% of the land area to be at risk of salination. With regards to agriculture, the World Bank targets food production and water management as an increasingly global issue that will foster debate.

Limited Resource

Fresh water is a renewable and variable, but finite natural resource. Fresh water can only be replenished through the process of the water cycle, in which water from seas, lakes, forests, land, rivers, and reservoirs evaporates, forms clouds, and returns as precipitation. Locally however, if more fresh water is consumed through human activities than is naturally restored, this may result in reduced fresh water availability from surface and underground sources and can cause serious damage to surrounding and associated environments.

Fresh Water Withdrawal

Fresh water withdrawal is the quantity of water removed from available sources for use in any purpose, excluding evaporation losses. Water drawn off is not necessarily entirely consumed and some portion may be returned for further use downstream.

Causes of Limited Fresh Water

There are many causes of the apparent decrease in our fresh water supply. Principal amongst these is the increase in population through increasing life expectancy, the increase in per capita water use and the desire of many people to live in warm climates that have naturally low levels of fresh water resources. Climate change is also likely to change the availability and distribution of fresh water across the planet:

"If global warming continues to melt glaciers in the polar regions, as expected, the supply of fresh water may actually decrease. First, fresh water from the melting glaciers will mingle with salt water in the oceans and become too salty to drink. Second, the increased ocean volume will cause sea levels to rise, contaminating freshwater sources along coastal regions with seawater".

The World Bank adds that the response by freshwater ecosystems to a changing climate can be described in terms of three interrelated components: water quality, water quantity or volume, and water timing. A change in one often leads to shifts in the others as well. Water pollution and subsequent eutrophication also reduces the availability of fresh water.

Also, there is an uneven distribution of fresh water. While some countries have an abundant supply of fresh water, others do not have as much. For example, Canada has 20% of the world's fresh water supply, while India has only 10% of the world's fresh water supply, even though India's population is more than 30 times larger than that of Canada. A reason for the uneven distribution of fresh water supply may be the differences in climate. For example, in some countries in Africa, the frequent lack of rain has led to insufficient water supply for irrigation. This has affected agriculture and has led to a shortage of food for the people.

Fresh Water in the Future

Many areas of the world are already experiencing stress on water availability. Due to the accelerated pace of population growth and an increase in the amount of water a single person uses, it is expected that this situation will continue to get worse. A shortage of water in the future would be detrimental to the human population as it would affect everything from sanitation, to overall health and the production of grain.

Choices in the use of Fresh Water

With one in eight people in the world not having access to safe water it is important to use this resource in a prudent manner. Making the best use of water on a local basis probably provides the best solution. Local communities need to plan their use of fresh water and should be made aware of how certain crops and animals use water.

As a guide the following tables provide some indicators.

Table 1 Recommended basic water requirements for human needs (per person)

Activity	Minimum, litres / day	Range / day
Drinking Water	5	2–5
Sanitation Services	20	20–75
Bathing	15	5–70
Cooking and Kitchen	10	10–50

Table 2. Water Requirements of different classes of livestock

Animal	Average / day	Range / day
Dairy cow	76 L (20 US gal)	57 to 95 L (15 to 25 US gal)
Cow-calf pair	57 L (15 US gal)	8 to 76 L (2 to 20 US gal)
Yearling cattle	38 L (10 US gal)	23 to 53 L (6 to 14 US gal)
Horse	38 L (10 US gal)	30 to 53 L (8 to 14 US gal)
Sheep	8 L (2 US gal)	8 to 11 L (2 to 3 US gal)

Table 3 Approximate values of seasonal crop water needs

Crop	Crop water needs mm / total growing period
Banana	1200–2200
Barley/Oats/Wheat	450–650
Cabbage	350–500
Citrus	900–1200
Onions	350–550
Pea	350–500
Potato	500–700
Sugar Cane	1500–2500
Tomato	400–800

Accessing Fresh Water

Canada

Canada has approximately 7% of the world's renewable fresh water. Canadians access their water from ground water, lakes and streams; it is then cleaned and purified in water treatment plants.

United States

The United States uses much more water per capita than developing countries. For example, the average American's daily shower uses more water than a person in a developing country would use for an entire day. Las Vegas, a city that uses an extreme amount of water to support spectacular lush greenery and golf courses, as well as huge fountains and swimming pools gets 90% of its water from Lake Mead, which is now at a record all-time low.

Developing Countries

In developing countries, 780 million people lack access to clean water. Half of the population of the developing world suffers from at least one disease caused by insufficient water supply and sanitation.

References

- Metzger, Bruce M.; Coogan, Michael D. (1993). The Oxford Companion to the Bible. New York, NY: Oxford University Press. p. 369. ISBN 0195046455.

- Merrill, Eugene H.; Rooker, Mark F.; Grisanti, Michael A. (2011). The World and the Word. Nashville, TN: B&H Academic. p. 430. ISBN 9780805440317.

- Franson, Mary Ann (1975). Standard Methods for the Examination of Water and Wastewater 14th ed. Washington, DC: American Public Health Association, American Water Works Association & Water Pollution Control Federation. ISBN 0-87553-078-8

- Peter Gleick; Heather Cooley; David Katz (2006). The world's water, 2006–2007: the biennial report on freshwater resources. Island Press. pp. 29–31. ISBN 1-59726-106-8. Retrieved 12 September 2009.

- "Center for Coastal Monitoring and Assessment: Mussel Watch Contaminant Monitoring". Ccma.nos.noaa.gov. 2014-01-14. Retrieved 2015-09-04.

- "Earth found hiding huge reservoirs of water 400 miles below...but not water as we know it : SCIENCE : Tech Times". Tech Times. 16 June 1015. Retrieved 13 October 2015.

- Becky Oskin (12 Mar 2014). "Rare Diamond Confirms That Earth's Mantle Holds an Ocean's Worth of Water". Scientific American. Retrieved 13 Mar 2015.

- Gleeson, Tom; Wada, Yoshihide; Bierkens, Marc F. P.; van Beek, Ludovicus P. H. (9 August 2012). "Water balance of global aquifers revealed by groundwater footprint". Nature. 488 (7410): 197–200. doi:10.1038/nature11295. Retrieved 2013-05-29.

- Justin Gillis (April 26, 2012). "Study Indicates a Greater Threat of Extreme Weather". The New York Times. Retrieved April 27, 2012.

- "LUHNA Chapter 6: Historical Landcover Changes in the Great Lakes Region". Biology.usgs.gov. 2003-11-20. Archived from the original on 2012-01-11. Retrieved 2011-02-19.

- The World Bank, 2009 "Water and Climate Change: Understanding the Risks and Making Climate-Smart Investment Decisions". pp. 19–22. Retrieved 24 October 2011.

- The World Bank, 2010 "Sustaining water for all in a changing climate: World Bank Group Implementation Progress Report". Retrieved 2011-10-24.

- The World Bank, 2009 "Water and Climate Change: Understanding the Risks and Making Climate-Smart Investment Decisions". pp. xv. Retrieved 2011-10-24.

3

Challenges and Concerns of Water Conservation

The concerns of water conservation elucidated are water scarcity, water security, peak water, waste-water, etc. Water scarcity involves water shortage, water stress and water crisis while wastewater is any water that has been adversely affected by human influence. Conservation of water should be of utmost importance, not only for the present generation but also for the future generation.

Water Scarcity

Water crisis is the lack of sufficient available water resources to meet water needs within a region. It affects every continent and around 2.8 billion people around the world at least one month out of every year. More than 1.2 billion people lack access to clean drinking water.

In Meatu district, Simiyu Region, Tanzania (Africa), water most often comes from open holes dug in the sand of dry riverbeds, and it is invariably contaminated.

Water scarcity involves water shortage, water stress or deficits, and water crisis. The relatively new concept of *water stress* is difficulty in obtaining sources of fresh water for use during a period of time; it may result in further depletion and deterioration of available water resources. *Water shortages* may be caused by climate change, such as altered weather-patterns (including droughts or floods), increased pollution, and increased human demand and overuse of water. The term *water crisis* labels a situation where the available potable, unpolluted water within a region is less than that region's demand. Two converging phenomena drive water scarcity: growing freshwater use and depletion of usable freshwater resources.

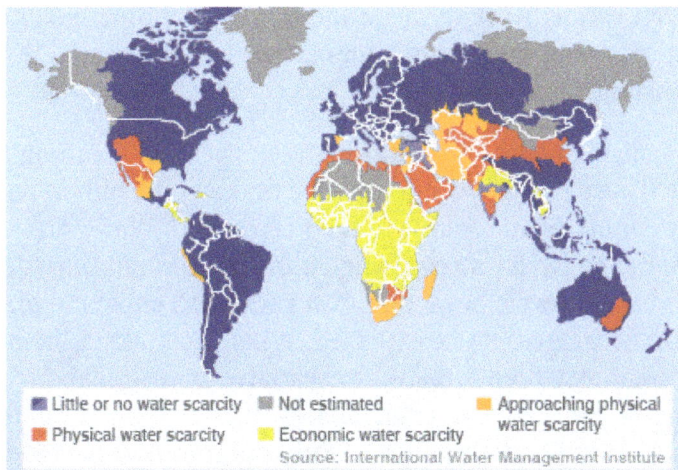

Physical water scarcity and economic water scarcity by country. 2006

Water scarcity can result from two mechanisms:

- physical (absolute) water scarcity

- economic water scarcity

Physical water scarcity results from inadequate natural water resources to supply a region's demand, and economic water scarcity results from poor management of the sufficient available water resources. According to the United Nations Development Programme, the latter is found more often to be the cause of countries or regions experiencing water scarcity, as most countries or regions have enough water to meet household, industrial, agricultural, and environmental needs, but lack the means to provide it in an accessible manner.

Many countries and governments aim to reduce water scarcity. The UN recognizes the importance of reducing the number of people without sustainable access to clean water and sanitation. The Millennium Development Goals within the United Nations Millennium Declaration aimed by 2015 to "halve the proportion of people who are unable to reach or to afford safe drinking water."

Water Stress

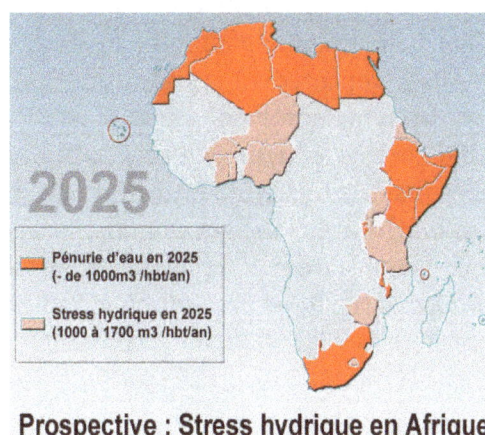

Prospective : Stress hydrique en Afrique

NGO estimate for 2025, 25 African countries are expected to suffer from water shortage or water stress.

The United Nations (UN) estimates that, of 1.4 billion cubic kilometers (1 quadrillion acre-feet) of water on Earth, just 200,000 cubic kilometers (162.1 billion acre-feet) represent fresh water available for human consumption.

More than one in every six people in the world is water stressed, meaning that they do not have access to potable water. Those that are water stressed make up 1.1 billion people in the world and are living in developing countries. According to the Falkenmark Water Stress Indicator, a country or region is said to experience "water stress" when annual water supplies drop below 1,700 cubic metres per person per year. At levels between 1,700 and 1,000 cubic meters per person per year, periodic or limited water shortages can be expected. When a country is below 1,000 cubic meters per person per year, the country then faces water scarcity . In 2006, about 700 million people in 43 countries were living below the 1,700 cubic metres per person threshold. Water stress is ever intensifying in regions such as China, India, and Sub-Saharan Africa, which contains the largest number of water stressed countries of any region with almost one fourth of the population living in a water stressed country. The world's most water stressed region is the Middle East with averages of 1,200 cubic metres of water per person. In China, more than 538 million people are living in a water-stressed region. Much of the water stressed population currently live in river basins where the usage of water resources greatly exceed the renewal of the water source.

Changes in Climate

Another popular opinion is that the amount of available freshwater is decreasing because of climate change. Climate change has caused receding glaciers, reduced stream and river flow, and shrinking lakes and ponds. Many aquifers have been over-pumped and are not recharging quickly. Although the total fresh water supply is not used up, much has become polluted, salted, unsuitable or otherwise unavailable for drinking, industry and agriculture. To avoid a global water crisis, farmers will have to strive to increase productivity to meet growing demands for food, while industry and cities find ways to use water more efficiently.

A New York Times article, "Southeast Drought Study Ties Water Shortage to Population, Not Global Warming", summarizes the findings of Columbia University researcher on the subject of the droughts in the American Southeast between 2005 and 2007. The findings published in the *Journal of Climate* say that the water shortages resulted from population size more than rainfall. Census figures show that Georgia's population rose from 6.48 to 9.54 million between 1990 and 2007. After studying data from weather instruments, computer models, and tree ring measurements, they found that the droughts were not unprecedented and result from normal climate patterns and random weather events. "Similar droughts unfolded over the last thousand years", the researchers wrote, "Regardless of climate change, they added, similar weather patterns can be expected regularly in the future, with similar results." As the temperature increases, rainfall in the Southeast will increase but because of evaporation the area may get even drier. The researchers concluded with a statement saying that any rainfall comes from complicated internal processes in the atmosphere and are very hard to predict because of the large amount of variables.

Water Crisis

When there is not enough potable water for a given population, the threat of a *water crisis* is realized. The United Nations and other world organizations consider a variety of regions to have water crises of global concern. Other organizations, such as the Food and Agriculture Organization, argue that there are no water crises in such places, but steps must still be taken to avoid one.

Effects of Water Crisis

There are several principal manifestations of the water crisis.

- Inadequate access to safe drinking water for about 884 million people

- Inadequate access to sanitation for 2.5 billion people, which often leads to water pollution

- Groundwater overdrafting (excessive use) leading to diminished agricultural yields

- Overuse and pollution of water resources harming biodiversity

- Regional conflicts over scarce water resources sometimes resulting in warfare.

Waterborne diseases caused by lack of sanitation and hygiene are one of the leading causes of death worldwide. For children under age five, waterborne diseases are a leading cause of death. According to the World Bank, 88 percent of all waterborne diseases are caused by unsafe drinking water, inadequate sanitation and poor hygiene.

Water is the underlying tenuous balance of safe water supply, but controllable factors such as the management and distribution of the water supply itself contribute to further scarcity.

A 2006 United Nations report focuses on issues of governance as the core of the water crisis, saying "There is enough water for everyone" and "Water insufficiency is often due to mismanagement, corruption, lack of appropriate institutions, bureaucratic inertia and a shortage of investment in both human capacity and physical infrastructure". Official data also shows a clear correlation between access to safe water and GDP per capita.

It has also been claimed, primarily by economists, that the water situation has occurred because of a lack of property rights, government regulations and subsidies in the water sector, causing prices to be too low and consumption too high.

Deforestation of the Madagascar Highland Plateau has led to extensive
siltation and unstable flows of western rivers.

Vegetation and wildlife are fundamentally dependent upon adequate freshwater resources. Marshes, bogs and riparian zones are more obviously dependent upon sustainable water supply, but forests and other upland ecosystems are equally at risk of significant productivity changes as water availability is diminished. In the case of wetlands, considerable area has been simply taken from wildlife use to feed and house the expanding human population. But other areas have suffered reduced productivity from gradual diminishing of freshwater inflow, as upstream sources are diverted for human use. In seven states of the U.S. over 80 percent of all historic wetlands were filled by the 1980s, when Congress acted to create a "no net loss" of wetlands.

In Europe extensive loss of wetlands has also occurred with resulting loss of biodiversity. For example, many bogs in Scotland have been developed or diminished through human population expansion. One example is the Portlethen Moss in Aberdeenshire.

On Madagascar's highland plateau, a massive transformation occurred that eliminated virtually all the heavily forested vegetation in the period 1970 to 2000. The slash and burn agriculture eliminated about ten percent of the total country's native biomass and converted it to a barren wasteland. These effects were from overpopulation and the necessity to feed poor indigenous peoples, but the adverse effects included widespread gully erosion that in turn produced heavily silted rivers that "run red" decades after the deforestation. This eliminated a large amount of usable fresh water and also destroyed much of the riverine ecosystems of several large west-flowing rivers. Several fish species have been driven to the edge of extinction and some, such as the disturbed Tokios coral reef formations in the Indian Ocean, are effectively lost. In October 2008, Peter Brabeck-Letmathe, chairman and former chief executive of Nestlé, warned that the production of biofuels will further deplete the world's water supply.

Overview of Regions Suffering Crisis Impacts

There are many other countries of the world that are severely impacted with regard to human health and inadequate drinking water. The following is a partial list of some of the countries with significant populations (numerical population of affected population listed) whose only consumption is of contaminated water:

- Sudan 12.3 million

- Venezuela 5.0 million

- Ethiopia 2.7 million

- Tunisia 2.1 million

- Cuba 1.3 million

According to the California Department of Resources, if more supplies aren't found by 2020, the region will face a shortfall nearly as great as the amount consumed today. Los Angeles is a coastal desert able to support at most 1 million people on its own water; the Los Angeles basin now is the core of a megacity that spans 220 miles (350 km) from Santa Barbara to the Mexican border. The region's population is expected to reach 41 million by 2020, up from 28 million in 2009. The population of California continues to grow by more than two million a year and is expected to reach 75 million in 2030, up from 49 million in 2009. But water shortage is likely to surface well before then.

Water deficits, which are already spurring heavy grain imports in numerous smaller countries, may soon do the same in larger countries, such as China and India. The water tables are falling in scores of countries (including Northern China, the US, and India) due to widespread overpumping using powerful diesel and electric pumps. Other countries affected include Pakistan, Iran, and Mexico. This will eventually lead to water scarcity and cutbacks in grain harvest. Even with the overpumping of its aquifers, China is developing a grain deficit. When this happens, it will almost certainly drive grain prices upward. Most of the 3 billion people projected to be added worldwide by mid-century will be born in countries already experiencing water shortages. Unless population growth can be slowed quickly, it is feared that there may not be a practical non-violent or humane solution to the emerging world water shortage.

After China and India, there is a second tier of smaller countries with large water deficits — Algeria, Egypt, Iran, Mexico, and Pakistan. Four of these already import a large share of their grain. But with a population expanding by 4 million a year, they will also likely soon turn to the world market for grain.

According to a UN climate report, the Himalayan glaciers that are the sources of Asia's biggest rivers – Ganges, Indus, Brahmaputra, Yangtze, Mekong, Salween and Yellow – could disappear by 2035 as temperatures rise. It was later revealed that the source used by the UN climate report actually stated 2350, not 2035. Approximately 2.4 billion people live in the drainage basin of the Himalayan rivers. India, China, Pakistan, Bangladesh, Nepal and Myanmar could experience floods followed by droughts in coming decades. In India alone, the Ganges provides water for drinking and farming for more than 500 million people. The west coast of North America, which gets much of its water from glaciers in mountain ranges such as the Rocky Mountains and Sierra Nevada, also would be affected.

By far the largest part of Australia is desert or semi-arid lands commonly known as the outback. In June 2008 it became known that an expert panel had warned of long term, possibly irreversible, severe ecological damage for the whole Murray-Darling basin if it does not receive sufficient water by October. Water restrictions are currently in place in many regions and cities of Australia in response to chronic shortages resulting from drought. The Australian of the year 2007, environmentalist Tim Flannery, predicted that unless it made drastic changes, Perth in Western Australia could become the world's first ghost metropolis, an abandoned city with no more water to sustain its population. However, Western Australia's dams reached 50% capacity for the first time since 2000 as of September 2009. As a result, heavy rains have brought forth positive results for the region. Nonetheless, the following year, 2010, Perth suffered its second-driest winter on record and the water corporation tightened water restrictions for spring.

Physical and Economic Scarcity

Around one fifth of the world's population currently live in regions affected by Physical water scarcity, where there is inadequate water resources to meet a country's or regional demand, including the water needed to fulfill the demand of ecosystems to function effectively. Arid regions frequently suffer from physical water scarcity. It also occurs where water seems abundant but where resources are over-committed, such as when there is over development of hydraulic infrastructure for irrigation. Symptoms of physical water scarcity include environmental degradation and declining groundwater as well as other forms of exploitation or overuse.

Economic water scarcity is caused by a lack of investment in infrastructure or technology to draw water from rivers, aquifers or other water sources, or insufficient human capacity to satisfy the demand for water. One quarter of the world's population is affected by economic water scarcity. Economic water scarcity includes a lack of infrastructure, causing the people without reliable access to water to have to travel long distances to fetch water, that is often contaminated from rivers for domestic and agricultural uses. Large parts of Africa suffer from economic water scarcity; developing water infrastructure in those areas could therefore help to reduce poverty. Critical conditions often arise for economically poor and politically weak communities living in already dry environment. Consumption increases with GDP per capita in most developed countries the average amount is around 200-300 litres daily. In underdeveloped (e.g. African countries such as Mozambique, average daily water consumption per capita was below 10 l. This is against the backdrop of international organisations, which recommend a minimum of 20 l of water (not including the water needed for washing clothes), available at most 1 km from the household. Increased water consumption is correlated with increasing income, as measured by GDP per capita. In countries suffering from water shortages water is the subject of speculation.

Human Right to Water

The United Nations Committee on Economic, Social and Cultural Rights established a foundation of five core attributes for water security. They declare that the human right to water entitles everyone to sufficient, safe, acceptable, physically accessible, and affordable water for personal and domestic use.

Millennium Development Goals (MDG)

At the 2000 Millennium Summit, the United Nations addressed the effects of economic water scarcity by making increased access to safe drinking water an international development goal. During this time, they drafted the Millennium Development Goals and all 189 UN members agreed on eight goals. MDG 7 sets a target for reducing the proportion of the population without sustainable safe drinking water access by half by 2015. This would mean that more than 600 million people would gain access to a safe source of drinking water. In 2016, the Sustainable Development Goals replace the Millennium Development Goals.

Water Scarcity's Effects on Environment

Water scarcity has many negative impacts on the environment, including lakes, rivers, wetlands, and other fresh water resources. The resulting water overuse that is related to water scarcity, often located in areas of irrigation agriculture, harms the environment in several ways including increased salinity, nutrient pollution, and the loss of floodplains and wetlands. Furthermore, water scarcity makes flow management in the rehabilitation of urban streams problematic.

Through the last hundred years, more than half of the Earth's wetlands have been destroyed and have disappeared. These wetlands are important not only because they are the habitats of numerous inhabitants such as mammals, birds, fish, amphibians, and invertebrates, but they support the growing of rice and other food crops as well as provide water filtration and protection from storms and flooding. Freshwater lakes such as the Aral Sea in central Asia have also suffered. Once the fourth largest freshwater lake, it has lost more than 58,000 square km of area and vastly increased in salt concentration over the span of three decades.

An abandoned ship in the former Aral Sea, near Aral, Kazakhstan.

Subsidence, or the gradual sinking of landforms, is another result of water scarcity. The U.S. Geological Survey estimates that subsidence has affected more than 17,000 square miles in 45 U.S. states, 80 percent of it due to groundwater usage. In some areas east of Houston, Texas the land has dropped by more than nine feet due to subsidence. Brownwood, a subdivision near Baytown, Texas, was abandoned due to frequent flooding caused by subsidence and has since become part of the Baytown Nature Center.

Climate Change

Aquifer drawdown or overdrafting and the pumping of fossil water increases the total amount of water within the hydrosphere subject to transpiration and evaporation processes, thereby causing accretion in water vapour and cloud cover, the primary absorbers of infrared radiation in the earth's atmosphere. Adding water to the system has a forcing effect on the whole earth system, an accurate estimate of which hydrogeological fact is yet to be quantified.

Depletion of Freshwater Resources

Apart from the conventional surface water sources of freshwater such as rivers and lakes, other resources of freshwater such as groundwater and glaciers have become more developed sources of freshwater, becoming the main source of clean water. Groundwater is water that has pooled below the surface of the Earth and can provide a usable quantity of water through springs or wells. These areas where groundwater is collected are also known as aquifers. Glaciers provide freshwater in the form meltwater, or freshwater melted from snow or ice, that supply streams or springs as temperatures rise. More and more of these sources are being drawn upon as conventional sources' usability decreases due to factors such as pollution or disappearance due to climate changes. The exponential growth rate of the human population is a main contributing factor in the increasing use of these types of water resources.

Groundwater

Until recent history, groundwater was not a highly utilized resource. In the 1960s, more and more groundwater aquifers developed. Changes in knowledge, technology and funding have allowed for focused development into abstracting water from groundwater resources away from surface water resources. These changes allowed for progress in society such as the "agricultural groundwater revolution," expanding the irrigation sector allowing for increased food production and development in rural areas. Groundwater supplies nearly half of all drinking water in the world. The large volumes of water stored underground in most aquifers have a considerable buffer capacity allowing for water to be withdrawn during periods of drought or little rainfall. This is crucial for people that live in regions that cannot depend on precipitation or surface water as a supply alone, instead providing reliable access to water all year round. As of 2010, the world's aggregated groundwater abstraction is estimated at approximately 1,000 km^3 per year, with 67% used for irrigation, 22% used for domestic purposes and 11% used for industrial purposes. The top ten major consumers of abstracted water (India, China, United States of America, Pakistan, Iran, Bangladesh, Mexico, Saudi Arabia, Indonesia, and Italy) make up 72% of all abstracted water use worldwide. Groundwater has become crucial for the livelihoods and food security of 1.2 to 1.5 billion rural households in the poorer regions of Africa and Asia.

Although groundwater sources are quite prevalent, one major area of concern is the renewal rate or recharge rate of some groundwater sources. Abstracting from groundwater sources that are non-renewable could lead to exhaustion if not properly monitored and managed. Another concern of increased groundwater usage is the diminished water quality of the source over time. Reduction of natural outflows, decreasing stored volumes, declining water levels and water degradation are commonly observed in groundwater systems. Groundwater depletion may result in many negative effects such as increased cost of groundwater pumping, induced salinity and other water quality changes, land subsidence, degraded springs and reduced baseflows. Human pollution is also harmful to this important resource.

Glaciers

Glaciers are noted as a vital water source due to their contribution to stream flow. Rising global temperatures have noticeable effects on the rate at which glaciers melt, causing glaciers in general to shrink worldwide. Although the meltwater from these glaciers are increasing the total water supply for the present, the disappearance of glaciers in the long term will diminish available water resources. Increased meltwater due to rising global temperatures can also have negative effects such as flooding of lakes and dams and catastrophic results.

Measurement of Water Scarcity

Hydrologists today typically assess water scarcity by looking at the population-water equation. This is done by comparing the amount of total available water resources per year to the population of a country or region. A popular approach to measuring water scarcity has been to rank countries according to the amount of annual water resources available per person. For example, according to the Falkenmark Water Stress Indicator, a country or region is said to experience "water stress" when annual water supplies drop below 1,700 cubic metres per person per year. At levels between

1,700 and 1,000 cubic metres per person per year, periodic or limited water shortages can be expected. When water supplies drop below 1,000 cubic metres per person per year, the country faces "water scarcity". The United Nations' FAO states that by 2025, 1.9 billion people will live in countries or regions with absolute water scarcity, and two-thirds of the world population could be under stress conditions. The World Bank adds that climate change could profoundly alter future patterns of both water availability and use,thereby increasing levels of water stress and insecurity, both at the global scale and in sectors that depend on water.

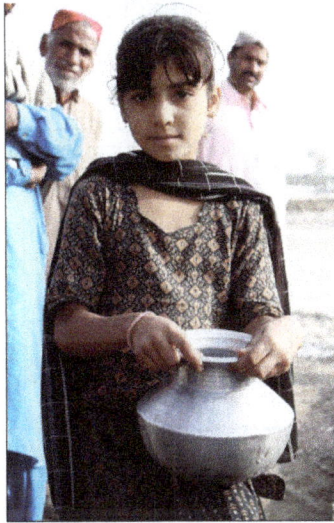

In 2012 in Sindh, Pakistan a shortage of clean water led people to queue to collect it where available

Other ways of measuring water scarcity include examining the physical existence of water in nature, comparing nations with lower or higher volumes of water available for use. This method often fails to capture the accessibility of the water resource to the population that may need it. Others have related water availability to population.

Another measurement, calculated as part of a wider assessment of water management in 2007, aimed to relate water availability to how the resource was actually used. It therefore divided water scarcity into 'physical' and 'economic'. Physical water scarcity is where there is not enough water to meet all demands, including that needed for ecosystems to function effectively. Arid regions frequently suffer from physical water scarcity. It also occurs where water seems abundant but where resources are over-committed, such as when there is overdevelopment of hydraulic infrastructure for irrigation. Symptoms of physical water scarcity include environmental degradation and declining groundwater. Water stress harms living things because every organism needs water to live.

Renewable Freshwater Resources

Renewable freshwater supply is a metric often used in conjunction when evaluating water scarcity. This metric is informative because it can describe the total available water resource each country contains. By knowing the total available water source, an idea can be gained about whether a country is prone to experiencing physical water scarcity. This metric has its faults in that it is an average; precipitation delivers water unevenly across the planet each year and annual renewable water resources vary from year to year. This metric also does not describe the accessibility of water to individuals, households, industries, or the government. Lastly, as this metric is a description of a whole country, it does not accurately portray whether a country is experiencing water scarcity. Canada and Brazil both have very high levels of available water supply, but still experience various water related problems.

It can be observed that tropical countries in Asia and Africa have low availability of freshwater resources.

The following table displays the average annual renewable freshwater supply by country including both surface-water and groundwater supplies. This table represents data from the UN FAO AQUASTAT, much of which are produced by modeling or estimation as opposed to actual measurements.

Total renewable freshwater supply by country				
Rank	Country	Annual renewable water resources (km³/year)	Region	Year of estimate
1	Kuwait	0.02	Asia	2008
2	St. Kitts and Nevis	0.02	North and Central America	2000
3	Maldives	0.03	Asia	1999
4	Malta	0.07	Europe	2005
5	Antigua and Barbuda	0.1	North and Central America	2000
6	Qatar	0.1	Asia	2008
7	Barbados	0.1	North and Central America	2003
8	Bahrain	0.1	Asia	2008
9	United Arab Emirates	0.2	Asia	2008
10	Cape Verde	0.3	Africa	2005
11	Djibouti	0.3	Africa	2005
12	Cyprus	0.3	Europe	2007
13	Libya	0.6	Africa	2005
14	Singapore	0.6	Asia	1975
15	Jordan	0.9	Asia	2008
16	Comoros	1.2	Africa	2005
17	Oman	1.4	Asia	2008
18	Luxembourg	1.6	Europe	2007
19	Israel	1.8	Asia	2008
20	Yemen	2.1	Asia	2008
21	Saudi Arabia	2.4	Asia	2008
22	Mauritius	2.8	Africa	2005
23	Burundi	3.6	Africa	1987
24	Trinidad and Tobago	3.8	North and Central America	2000

25	Swaziland	4.5	Africa	1987
26	Lebanon	4.5	Asia	2008
27	Tunisia	4.6	Africa	2005
28	Reunion	5.0	Africa	1988
29	Lesotho	5.2	Africa	1987
30	Eritrea	6.3	Africa	2001
31	Macedonia	6.4	Europe	2001
32	Armenia	7.8	Former Soviet Union	2008
33	Gambia	8.0	Africa	2005
34	Brunei	8.5	Asia	1999
35	Jamaica	9.4	North and Central America	2000
36	Rwanda	9.5	Africa	2005
37	Mauritania	11.4	Africa	2005
38	Algeria	11.6	Africa	2005
39	Moldova	11.7	Former Soviet Union	1997
40	Estonia	12.3	Europe	2007
41	Estonia	12.8	Former Soviet Union	1997
42	Haiti	14.0	North and Central America	2000
43	Somalia	14.2	Africa	2005
44	Botswana	14.7	Africa	2001
45	Togo	14.7	Africa	2001
46	Czech Republic	16.0	Europe	2007
47	Denmark	16.3	Europe	2007
48	Syria	16.8	Asia	2008
49	Malawi	17.3	Africa	2001
50	Burkina Faso	17.5	Africa	2001
51	Namibia	17.7	Africa	2005
52	Belize	18.6	North and Central America	2000
53	Zimbabwe	20.0	Africa	1987
54	Belgium	20.0	Europe	2007
55	Dominican Republic	21.0	North and Central America	2000
56	Lithuania	24.5	Former Soviet Union	2007
57	El Salvador	25.2	North and Central America	2001
58	Romania	25.7	Europe	2007
59	Benin	25.8	Africa	2001
60	Equatorial Guinea	26	Africa	2001
61	Fiji	28.6	Oceania	1987
62	Morocco	29.0	Africa	2005
63	Kenya	30.7	Africa	2005

64	Guinea-Bissau	31.0	Africa	2005
65	Slovenia	32.1	Europe	2007
66	Niger	33.7	Africa	2005
67	Azerbaijan	34.7	Former Soviet Union	2008
68	Mongolia	34.8	Asia	1999
69	Bosnia and Herzegovina	37.5	Europe	2003
70	Cuba	38.1	North and Central America	2000
71	Senegal	39.4	Africa	1987
72	Albania	41.7	Europe	2001
73	Chad	43.0	Africa	1987
74	Solomon Islands	44.7	Oceania	1987
75	Kyrgyzstan	46.5	Former Soviet Union	1997
76	Ireland	46.8	Europe	2003
77	South Africa	50.0	Africa	2005
78	Sri Lanka	50.0	Asia	1999
79	Slovakia	50.1	Europe	2007
80	Ghana	53.2	Africa	2001
81	Switzerland	53.5	Europe	2007
82	Belarus	58.0	Former Soviet Union	1997
83	Egypt	58.3	Africa	2005
84	Turkmenistan	60.9	Former Soviet Union	1997
85	Poland	63.1	Europe	2007
86	Georgia	63.3	Former Soviet Union	2008
87	Sudan	64.5	Africa	2005
88	Afghanistan	65.0	Asia	1997
89	Uganda	66.0	Africa	2005
90	Taiwan	67.0	Asia	2000
91	Korea Rep	69.7	Asia	1999
92	Greece	72.0	Europe	2007
93	Uzbekistan	72.2	Former Soviet Union	2003
94	Portugal	73.6	Europe	2007
95	Iraq	75.6	Asia	2008
96	Korea DPR	77.1	Asia	1999
97	Côte d'Ivoire	81	Africa	2001
98	Austria	84.0	Europe	2007
99	Netherlands	89.7	Europe	2007
100	Tanzania	91	Africa	2001
101	Bhutan	95.0	Asia	1987
102	Honduras	95.9	North and Central America	2000
103	Tajikistan	99.7	Former Soviet Union	1997
104	Mali	100.0	Africa	2005

105	Zambia	105.2	Africa	2001
106	Croatia	105.5	Europe	1998
107	Bulgaria	107.2	Europe	2010
108	Kazakhstan	109.6	Former Soviet Union	1997
109	Ethiopia	110.0	Africa	1987
110	Finland	110.0	Europe	2007
111	Spain	111.1	Europe	2007
112	Guatemala	111.3	North and Central America	2000
113	Costa Rica	112.4	North and Central America	2000
114	Hungary	116.4	Europe	2007
115	Suriname	122.0	South America	2003
116	Iran	137.5	Asia	2008
117	Uruguay	139.0	South America	2000
118	Ukraine	139.5	Former Soviet Union	1997
119	Central African Republic	144.4	Africa	2005
120	Panama	148.0	North and Central America	2000
121	Sierra Leone	160.0	Africa	1987
122	Gabon	164.0	Africa	1987
123	Iceland	170.0	Europe	2007
124	Italy	175.0	Europe	2007
125	United Kingdom	175.3	Europe	2007
126	Sweden	183.4	Europe	2007
127	Angola	184.0	Africa	1987
128	France	186.3	Europe	2007
129	Germany	188.0	Europe	2007
130	Nicaragua	196.7	North and Central America	2000
131	Serbia-Montenegro*	208.5	Europe	2003
132	Nepal	210.2	Asia	1999
133	Turkey	213.6	Asia	2008
134	Mozambique	217.1	Africa	2005
135	Guinea	226.0	Africa	1987
136	Liberia	232.0	Africa	1987
137	Pakistan	233.8	Asia	2003
138	Guyana	241.0	South America	2000
139	Cameroon	285.5	Africa	2003
140	Nigeria	286.2	Africa	2005
141	Laos	333.6	Asia	2003
142	Paraguay	336.0	South America	2000
143	Australia	336.1	Oceania	2005

144	Madagascar	337.0	Africa	2005
145	Latvia	337.3	Former Soviet Union	2007
146	Norway	389.4	Europe	2007
147	New Zealand	397.0	Oceania	1995
148	Thailand	409.9	Asia	1999
149	Japan	430.0	Asia	1999
150	Ecuador	432.0	South America	2000
151	Mexico	457.2	North and Central America	2000
152	Cambodia	476.1	Asia	1999
153	Philippines	479.0	Asia	1999
154	Malaysia	580.0	Asia	1999
155	Bolivia	622.5	South America	2000
156	Papua New Guinea	801.0	Oceania	1987
157	Argentina	814.0	South America	2000
158	Congo	832.0	Africa	1987
159	Vietnam	891.2	Asia	1999
160	Chile	922.0	South America	2000
161	Myanmar	1045.6	Asia	1999
162	Bangladesh	1210.6	Asia	1999
163	Venezuela	1233.2	South America	2000
164	Congo, Democratic Republic (formerly Zaire)	1283	Africa	2001
165	India	1907.8	Asia	1999
166	Peru	1913.0	South America	2000
167	Colombia	2132.0	South America	2000
168	China	2738.8	Asia	2008
169	Indonesia	2838.0	Asia	1999
170	United States of America	3069.0	North and Central America	1985
171	Canada	3300.0	North and Central America	1985
172	Russia	4498.0	Former Soviet Union	1997
173	Brazil	8233.0	South America	2000

Outlook

Construction of wastewater treatment plants and reduction of groundwater overdrafting appear to be obvious solutions to the worldwide problem; however, a deeper look reveals more fundamental issues in play. Wastewater treatment is highly capital intensive, restricting access to this technology in some regions; furthermore the rapid increase in population of many countries makes this a race that is difficult to win. As if those factors are not daunting enough, one must consider the enormous costs and skill sets involved to maintain wastewater treatment plants even if they are successfully developed.

Reducing groundwater overdrafting is usually politically unpopular, and can have major economic impacts on farmers. Moreover, this strategy necessarily reduces crop output, something the world can ill-afford given the current population.

Wind and solar power such as this installation in a village in northwest Madagascar can make a difference in safe water supply.

At more realistic levels, developing countries can strive to achieve primary wastewater treatment or secure septic systems, and carefully analyse wastewater outfall design to minimise impacts to drinking water and to ecosystems. Developed countries can not only share technology better, including cost-effective wastewater and water treatment systems but also in hydrological transport modeling. At the individual level, people in developed countries can look inward and reduce over-consumption, which further strains worldwide water consumption. Both developed and developing countries can increase protection of ecosystems, especially wetlands and riparian zones. These measures will not only conserve biota, but also render more effective the natural water cycle flushing and transport that make water systems more healthy for humans.

A range of local, low-tech solutions are being pursued by a number of companies. These efforts center around the use of solar power to distill water at temperatures slightly beneath that at which water boils. By developing the capability to purify any available water source, local business models could be built around the new technologies, accelerating their uptake. For example, bedouins from the town of Dahab in Egypt have installed AquaDania's WaterStillar, which uses a solar thermal collector measuring two square metres to distill from 40 to 60 litres per day from any local water source. This is five times more efficient than conventional stills and eliminates the need for polluting plastic PET bottles or transportation of water supply.

Global Experiences in Managing Water Crisis

It is alleged that the likelihood of conflict rises if the rate of change within the basin exceeds the capacity of institution to absorb that change. Although water crisis is closely related to regional tensions, history showed that acute conflicts over water are far less than the record of cooperation.

The key lies in strong institutions and cooperation. The Indus River Commission and the Indus Water Treaty survived two wars between India and Pakistan despite their hostility, proving to be a successful mechanism in resolving conflicts by providing a framework for consultation inspection and exchange of data. The Mekong Committee has also functioned since 1957 and survived the Vietnam War. In contrast, regional instability results when there is an absence of institutions to co-operate in regional collaboration, like Egypt's plan for a high dam on the Nile. However, there is currently no global institution in place for the management and management of trans-boundary water sources, and international co-operation has happened through ad hoc collaborations between agencies, like the Mekong Committee which was formed due to an alliance between UNICEF and the US Bureau of Reclamation. Formation of strong international institutions seems to be a way forward – they fuel early intervention and management, preventing the costly dispute resolution process.

One common feature of almost all resolved disputes is that the negotiations had a "need-based" instead of a "right–based" paradigm. Irrigable lands, population, technicalities of projects define "needs". The success of a need-based paradigm is reflected in the only water agreement ever negotiated in the Jordan River Basin, which focuses in needs not on rights of riparians. In the Indian subcontinent, irrigation requirements of Bangladesh determine water allocations of The Ganges River. A need based, regional approach focuses on satisfying individuals with their need of water, ensuring that minimum quantitative needs are being met. It removes the conflict that arises when countries view the treaty from a national interest point of view, move away from the zero-sum approach to a positive sum, integrative approach that equitably allocated the water and its benefits.

The Blue Peace framework developed by Strategic Foresight Group in partnership with the Governments of Switzerland and Sweden offers a unique policy structure which promotes sustainable management of water resources combined with cooperation for peace. By making the most of shared water resources through cooperation rather than mere allocation between countries, the chances for peace can be increased. The Blue Peace approach has proven to be effective in cases like the Middle East and the Nile basin.

Water Security

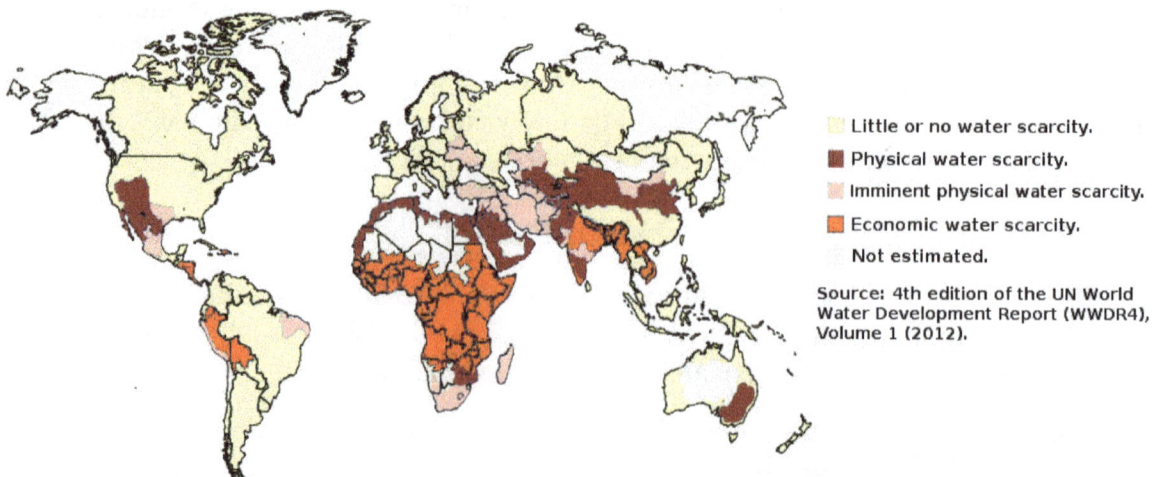

Little or no water scarcity.
Physical water scarcity.
Imminent physical water scarcity.
Economic water scarcity.
Not estimated.

Source: 4th edition of the UN World Water Development Report (WWDR4), Volume 1 (2012).

A 2012 map of the world showing in red the places with least water security.

Water security has been defined as "the reliable availability of an acceptable quantity and quality of water for health, livelihoods and production, coupled with an acceptable level of water-related risks." Sustainable development will not be achieved without a water secure world. A water secure world integrates a concern for the intrinsic value of water with a concern for its use for human survival and well-being. A water secure world harnesses water's productive power and minimises its destructive force. Water security also means addressing environmental protection and the negative effects of poor management. It is also concerned with ending fragmented responsibility for water and integrating water resources management across all sectors – finance, planning, agriculture, energy, tourism, industry, education and health. A water secure world reduces poverty, advances education, and increases living standards. It is a world where there is an improved quality of life for all, especially for the most vulnerable—usually women and children—who benefit most from good water governance.

Background

According to the Pacific Institute "While regional impacts will vary, global climate change will potentially alter agricultural productivity, freshwater availability and quality, access to vital minerals, coastal and island flooding, and more. Among the consequences of these impacts will be challenges to political relationships, realignment of energy markets and regional economies, and threats to security".

It impacts regions, states and countries. Tensions exist between upstream and downstream users of water within individual jurisdictions.

During history there has been much conflict over use of water from rivers such as the Tigris and Euphrates Rivers. Another highly politicized example is Israel's control of water resources in the Levant region since its creation, where Israel securing its water resources was one of several drivers for the 1967 Six Day War.

Water security is sometimes sought by implementing water desalination, pipelines between sources and users, water licences with different security levels and war.

Water allocation between competing users is increasingly determined by application of market-based pricing for either water licenses or actual water.

Fresh Water

Water, in absolute terms, is not in short supply planet-wide. But, according to the United Nations water organization, UN-Water, the total usable freshwater supply for ecosystems and humans is only about 200,000 km3 of water – less than one percent (<1%) of all freshwater resources. And, water use has been growing at more than twice the rate of the population increase in the last century. Specifically, water withdrawals are predicted to increase by 50 percent by 2025 in developing countries, and 18 per cent in developed countries. By 2025, 1.8 billion people will be living in countries or regions with absolute water scarcity, and two-thirds of the world population could be under stress conditions.

Research

According to *Nature* (2010), about 80% of the world's population (5.6 billion in 2011) live in areas with threats to water security. The water security is a shared threat to human and nature and it is pandemic. Human water-management strategies can be detrimental to wildlife, such as migrating fish. Regions with intensive agriculture and dense populations, such as the US and Europe, have a high threat of water security. Water is increasingly being used as a weapon in conflict. Water insecurity is always accompanied by one or more issues such as poverty, war and conflict, low women's development and environmental degradation. Researchers estimate that during 2010-2015, ca US$800 billion will be required to cover the annual global investment in water infrastructure. Good management of water resources can jointly manage biodiversity protection and human water security. Preserving flood plains rather than constructing flood-control reservoirs would provide a cost-effective way to control floods while protecting the biodiversity of wildlife that occupies such areas.

Thailand nan river fishing

Lawrence Smith, the president of the population institute, asserts that although an overwhelming majority of the planet's surface is composed of water, 97% of this water is constituted of saltwater; the fresh water used to sustain humans is only 3% of the total amount of water on Earth. Therefore, Smith believes that the competition for water in an overpopulated world would pose a major threat to human stability, even going so far as to postulate apocalyptic world wars being fought over the control of thinning ice sheets and nearly desiccated reservoirs. Nevertheless, 2 billion people have supposedly gained access to a safe water source since 1990 who may have earlier lacked it. The proportion of people in developing countries with access to safe water is calculated to have improved from 30 percent in 1970 to 71 percent in 1990, 79 percent in 2000 and 84 percent in 2004, parallel with rising population. This trend is projected to continue.

The Earth has a limited though renewable supply of fresh water, stored in aquifers, surface waters and the atmosphere. Oceans are a good source of usable water, but the amount of energy needed to convert saline water to potable water is prohibitive with conventional approaches, explaining why only a very small fraction of the world's water supply is derived from desalination. However, modern technologies, such as the Seawater Greenhouse, use solar energy to desalinate seawater for agriculture and drinking uses in an extremely cost-effective manner.

Most Affected Countries

Communal tap (standpost) for drinking water in Soweto, Johannesburg, South Africa

Based on the map published by the Consultative Group on International Agricultural Research (CGIAR), the countries and regions suffering most water stress are North Africa, the Middle East, India, Central Asia, China, Chile, Colombia, South Africa and Australia. Water scarcity is also increasing in South Asia.

International Competition

More than 50 countries on five continents are said to be at risk of conflict over water.

Turkey's Southeastern Anatolia Project (Guneydogu Anadolu Projesi, or GAP) on the Euphrates has potentially serious consequences for water supplies in Syria and Iraq.

China, is constructing dams on the Mekong, leaving Vietnam, Laos, Cambodia and Thailand without same amout of water as before investment. A huge project of reversing the flow of the Brahmaputra (Chinese: Tsangpo) river, which after leaving Chinese Tibet flows through India and Bangladesh. The struggle for water in some afflicted regions has led inhabitants to hiring guards in order to protect wells. Moreover, Amu Daria River, shared by Uzbekistan, Turkmenistan, Tajikistan and Afghanistan, which has been nearly completely dried out, so much so that it has ceased to reach the Aral Sea/Lake, which is evaporating in an alarming pace. The fact that Turkmenistan retains much of the water before it flows into Uzbekistan.

Intra-national Competition

Australia

In Australia there is competition for the resources of the Darling River system between Queensland, New South Wales and South Australia.

In Victoria, Australia a proposed pipeline from the Goulburn Valley to Melbourne has led to protests by farmers.

In the Macquarie Marshes of NSW grazing and irrigation interests compete for water flowing to the marshes

The Snowy Mountains Scheme diverted water from the Snowy River to the Murray River and the Murrumbidgee River for the benefit of irrigators and electricity generation through hydro-electric power. During recent years government has taken action to increase environmental flows to the Snowy in spite of severe drought in the Murray Darling Basin. The Australian Government has implemented buy-backs of water allocations, or properties with water allocations, to endeavour to increase environmental flows.

India

In India, there is competition for water resources of all inter state rivers except the main Brahmaputra river among the riparian states of India and also with neighbouring countries which are Nepal, China, Pakistan, Bhutan, Bangladesh, etc. Vast area of the Indian subcontinent is under tropical climate which is conducive for agriculture due to favourable warm and sunny conditions provided perennial water supply is available to cater to the high rate of evapotranspiration from the cultivated land. Though the overall water resources are adequate to meet all the requirements of the subcontinent, the water supply gaps due to temporal and spatial distribution of water resources among the states and countries in the subcontinent are to be bridged.

There is intense competition for the water available in the inter state rivers such as Kavery, Krishna, Godavari, Vamsadhara, Mandovi, Ravi-Beas-Sutlez, Narmada, Tapti, Mahanadi, etc. among the riparian states of India in the absence of water augmentation from the water surplus rivers such as Brahmaputra, Himalayan tributaries of Ganga and west flowing coastal rivers of western ghats. All river basins face severe water shortage even for drinking needs of people, cattle and wild life during the intense summer season when the rainfall is negligible.

Water security can be achieved along with energy security as it is going to consume electricity to link the surplus water areas with the water deficit areas by lift canals, pipe lines, etc. The total water resources going waste to the sea are nearly 1200 billion cubic meters after sparing moderate environmental / salt export water requirements of all rivers. Interlinking rivers of the subcontinent is possible to achieve water security in the Indian subcontinent with the active cooperation of the countries in the region.

Blue Peace

US Navy Seabees tapping an artesian well in Helmand Province, Afghanistan

Blue Peace is a method which seeks to transforms trans-boundary water issues into instruments for cooperation. This unique approach to turn tensions around water into opportunities for socio-economic development was developed by Strategic Foresight Group in partnership with the Governments of Switzerland and Sweden.

"The Blue Peace is an innovative approach to engage political leaders, diplomats and populations in harnessing and managing collaborative solutions for sustainable water management." – Foreign Minister Didier Burkhalter of Swirtzerland, speaking at the UN General Assembly

Conventional Fossil or Nuclear Energy Based Desalination

Due to record low rainfall in Summer 2005, the reservoir behind Sameura Dam runs low.
The reservoir supplies water to Takamatsu, Shikoku Island, Japan.

As new technological innovations continue to reduce the capital cost of desalination, more countries are building desalination plants as a small element in addressing their water crises.

- Israel desalinizes water for a cost of 53 cents per cubic meter

- Singapore desalinizes water for 49 cents per cubic meter and also treats sewage with reverse osmosis for industrial and potable use (NEWater).

- China and India, the world's two most populous countries, are turning to desalination to provide a small part of their water needs

- In 2007 Pakistan announced plans to use desalination

- All Australian capital cities (except Darwin, Northern Territory and Hobart) are either in the process of building desalination plants, or are already using them. In late 2011, Melbourne will begin using Australia's largest desalination plant, the Wonthaggi desalination plant to raise low reservoir levels.

- In 2007 Bermuda signed a contract to purchase a desalination plant

- The largest desalination plant in the United States is the one at Tampa Bay, Florida, which began desalinizing 25 million gallons (95000 m^3) of water per day in December 2007. In the United States, the cost of desalination is $3.06 for 1,000 gallons, or 81 cents per cubic meter. In the United States, California, Arizona, Texas, and Florida use desalination for a very small part of their water supply.

- After being desalinized at Jubail, Saudi Arabia, water is pumped 200 miles (320 km) inland though a pipeline to the capital city of Riyadh.

A January 17, 2008, article in the *Wall Street Journal* states, "World-wide, 13,080 desalination plants produce more than 12 billion gallons of water a day, according to the International Desalination Association."

The world's largest desalination plant is the Jebel Ali Desalination Plant (Phase 2) in the United Arab Emirates. It is a dual-purpose facility that uses multi-stage flash distillation and is capable of producing 300 million cubic meters of water per year.

A typical aircraft carrier in the U.S. military uses nuclear power to desalinize 400,000 US gallons (1,500,000 L) of water per day.

While desalinizing 1,000 US gallons (3,800 L) of water can cost as much as $3, the same amount of bottled water costs $7,945.

However, given the energy intensive nature of desalination, with associated economic and environmental costs, desalination is generally considered a last resort after water conservation. But this is changing as prices continue to fall.

According to MSNBC, a report by Lux Research estimated that the worldwide desalinated water supply will triple between 2008 and 2020.

However, not everyone is convinced that desalination is or will be economically viable or environmentally sustainable for the foreseeable future. Debbie Cook, the former mayor of Huntington Beach, California, has been a frequent critic of desalination proposals ever since she was appointed as a member of the California Desalination Task Force. Cook claims that in addition to being energy intensive, desalination schemes are very costly—often much more costly than desalination proponents claim. In her writing on the subject, Cook points to a long list of projects that have stalled or been aborted for financial or other reasons, and suggests that water-stressed regions would do better to focus on conservation or other water supply solutions than to invest in desalination plants.

Solar Energy Based Desalination

A novel approach to desalination is the Seawater Greenhouse which takes seawater and uses solar energy to desalinate it in conjunction with growing food crops in a specially adapted greenhouse.

Peak Water

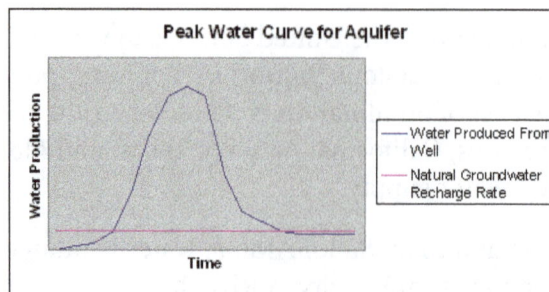

Potential peak water curve for production of groundwater from an aquifer.

Peak water is a concept that underlines the growing constraints on the availability, quality, and use of freshwater resources.

Peak water is defined in a 2010 peer-reviewed article in the *Proceedings of the National Academy of Sciences* by Peter Gleick and Meena Palaniappan. They distinguish between peak renewable, peak non-renewable, and peak ecological water in order to demonstrate the fact that although there is a vast amount of water on the planet, sustainably managed water is becoming scarce.

Lester R. Brown, president of the Earth Policy Institute, wrote in 2013 that although there was extensive literature on peak oil, it was peak water that is "the real threat to our future". An assessment was published in August 2011 in the Stockholm International Water Institute's journal. Much of the world's water in underground aquifers and in lakes can be depleted and thus resembles a finite resource. The phrase *peak water* sparks debates similar to those about peak oil. In 2010, *New York Times* chose "peak water" as one of its 33 "Words of the Year".

There are concerns about impending peak water in several areas around the world:

- Peak renewable water, where entire renewable flows are being consumed for human use

- Peak non-renewable water, where groundwater aquifers are being overpumped (or contaminated) faster than nature recharges them (this example is most like the peak oil debate)

- Peak ecological water, where ecological and environmental constraints are overwhelming the economic benefits provided by water use

If present trends continue, 1.8 billion people will be living with absolute water scarcity by 2025, and two-thirds of the world could be subject to water stress. Ultimately, peak water is not about running out of freshwater, but about reaching physical, economic, and environmental limits on meeting human demands for water and the subsequent decline of water availability and use.

Comparison with Peak Oil

The Hubbert curve has become popular in the scientific community for predicting the depletion of various natural resources. M. King Hubbert created this measurement device in 1956 for a variety of finite resources such as coal, oil, natural gas and uranium. Hubbert's curve was not applied to resources such as water originally, since water is a renewable resource. Some forms of water, however, such as fossil water, exhibit similar characteristics to oil, and overpumping (faster than the rate of natural recharge of groundwater) often results in a Hubbert-type peak. A modified Hubbert curve applies to any resource that can be harvested faster than it can be replaced. Like peak oil, peak water is inevitable given the rate of extraction of certain water systems. A current argument is that growing populations and demands for water will inevitably lead to non-renewable use of water resources.

Water Supply

Fresh water is a renewable resource, yet the world's supply of clean, fresh water is under increasing demand for human activities. The world has an estimated 1.34 billion cubic kilometers of water, but 96.5% of it is salty. Almost 70% of fresh water can be found in the ice caps of Antarctica and Greenland. Less than 1% of this water on Earth is accessible to humans, the rest is contained in soil moisture or deep underground. Accessible freshwater is located in lakes, rivers, reservoirs and shallow underground sources. Rainwater and snowfall do very little to replenish these stocks of freshwater.

Freshwater sources (top 15 countries)		
Country	Total freshwater supply	
	(km³/yr)	Year
Brazil	8233	2000
Russia	4498	1997
Canada	3300	1985
Colombia	3132	2000
USA	3069	1985
Indonesia	2838	1999
China	2830	2008
Peru	1913	2000
India	1908	1999
DR Congo	1283	2001
Venezuela	1233	2000
Bangladesh	1211	1999
Burma	1046	1999
Chile	922	2000
Vietnam	891	1999

The amount of available freshwater supply in some regions is decreasing because of (i) climate change, which has caused receding glaciers, reduced stream and river flow, and shrinking lakes; (ii) contamination of water by human and industrial wastes; and (iii) overuse of non-renewable groundwater aquifers. Many aquifers have been over-pumped and are not recharging quickly. Although the total freshwater supply is not used up, much has become polluted, salted, unsuitable or otherwise unavailable for drinking, industry and agriculture.

Water Demand

Water demand already exceeds supply in many parts of the world, and as the world population continues to rise, many more areas are expected to experience this imbalance in the near future.

Agriculture represents 70% of freshwater use worldwide.

Agriculture, industrialization and urbanization all serve to increase water consumption.

Freshwater Withdrawal by Country

The highest total annual consumption of water comes from India, China and the United States, countries with large populations, extensive agricultural irrigation, and demand for food. See the following table:

Freshwater withdrawal by country and sector (top 20 countries)					
Country	Total freshwater withdrawal (km³/yr)	Per capita withdrawal (m³/p/yr)	Domestic use (m³/p/yr)(in %)	Industrial use (m³/p/yr)(in %)	Agricultural use (m³/p/yr)(in %)
India	645.84	585	47 (8%)	30 (5%)	508 (86%)
China	549.76	415	29 (7%)	107 (26%)	279 (68%)
United States	477	1,600	208 (13%)	736 (46%)	656 (41%)
Pakistan	169.39	1,072	21 (2%)	21 (2%)	1029 (96%)
Japan	88.43	690	138 (20%)	124 (18%)	428 (62%)
Indonesia	82.78	372	30 (8%)	4 (1%)	339 (91%)
Thailand	82.75	1,288	26 (2%)	26 (2%)	1236 (95%)
Bangladesh	79.4	560	17 (3%)	6 (1%)	536 (96%)
Mexico	78.22	731	126 (17%)	37 (5%)	569 (77%)
Russia	76.68	535	102 (19%)	337 (63%)	96 (18%)
Iran	72.88	1,048	73 (7%)	21 (2%)	954 (91%)
Vietnam	71.39	847	68 (8%)	203 (24%)	576 (68%)
Egypt	68.3	923	74 (8%)	55 (6%)	794 (86%)
Brazil	59.3	318	64 (20%)	57 (18%)	197 (62%)
Uzbekistan	58.34	2,194	110 (5%)	44 (2%)	2040 (93%)
Canada	44.72	1,386	274 (20%)	947 (69%)	165 (12%)
Iraq	42.7	1,482	44 (3%)	74 (5%)	1363 (92%)
Italy	41.98	723	130 (18%)	268 (37%)	325 (45%)
Turkey	39.78	544	82 (15%)	60 (11%)	403 (74%)
Germany	38.01	460	55 (12%)	313 (68%)	92 (20%)

India

Working rice paddies

India has 20 percent of the Earth's population, but only four percent of its water. Water tables are dropping rapidly in some of India's main agricultural areas. The Indus and Ganges rivers are tapped so heavily that, except in rare wet years, they no longer reach the sea.

India has the largest water withdrawal out of all the countries in the world. Eighty-six percent of that water supports agriculture. That heavy use is dictated in large part by what people eat. People in India consume a lot of rice. Rice farmers in India typically get less than half the yield per unit area while using ten times more water than their Chinese counterparts. Economic development can make things worse because as people's living standards rise, they tend to eat more meat, which requires lots of water to produce. Growing a tonne of grain requires 1,000 tonnes of water; producing a tonne of beef requires 15,000 tonnes. To make a single hamburger requires around 4,940 liters (1,300 gallons) of water A glass of orange juice needs 850 liters (225 gallons) of freshwater to produce.

China

China, the world's most populous country, has the second largest water withdrawal; 68% supports agriculture while its growing industrial base consumes 26%. China is facing a water crisis where water resources are overallocated, used inefficiently, and severely polluted by human and industrial wastes. One-third of China's population lacks access to safe drinking water. Rivers and lakes are dead and dying, groundwater aquifers are over-pumped, uncounted species of aquatic life have been driven to extinction, and direct adverse impacts on both human and ecosystem health are widespread and growing.

In western China's Qinghai province, through which the Yellow River's main stream flows, more than 2,000 lakes have disappeared over the last 20 years. There were once 4,077 lakes. Global climate change is responsible for the reduction in flow of the (Huang He) Yellow River over the past several decades. The source of the Yellow River is the Qinghai-Xizang Tibetan Plateau where the glaciers are receding sharply.

In Hebei Province, which surrounds Beijing, the situation is much worse. Hebei is one of China's major wheat and corn growing provinces. The water tables have been falling fast throughout Hebei. The region has lost 969 of its 1,052 lakes. About 500,000 people are affected by a shortage of drinking water due to continuing droughts. Hydro-power generation is also impacted. Beijing and Tianjin depend on Hebei Province to supply their water from the Yangtze River. Beijing gets its water via the newly constructed South-North Water Transfer Project. The river originates in a glacier on the eastern part of the Tibetan Plateau.

United States

The United States has about 5% of the world's population, yet it uses almost as much water as India (~1/5 of world) or China (1/5 of world) because substantial amounts of water are used to grow food exported to the rest of the world. The United States agricultural sector consumes more water than the industrial sector, though substantial quantities of water are withdrawn (but not consumed) for power plant cooling systems. 40 out of 50 state water managers expect some degree of water stress in their state in the next 10 years.

The Ogallala Aquifer in the southern high plains (Texas and New Mexico) is being mined at a rate that far exceeds replenishment—a classic example of peak non-renewable water. Portions of the aquifer will not naturally recharge due to layers of clay between the surface and the water-bearing formation, and because rainfall rates simply do not match rates of extraction for irrigation. The term fossil water is sometimes used to describe water in aquifers that was stored over centuries to millennia. Use of this water is not sustainable when the recharge rate is slower than the rate of groundwater extraction.

Ship canal terminus

In California, massive amounts of groundwater are also being withdrawn from Central Valley groundwater aquifers—unreported, unmonitored, and unregulated. California's Central Valley is home to one-sixth of all irrigated land in the United States, and the state leads the nation in agricultural production and exports. The inability to sustain groundwater withdrawals over time may lead to adverse impacts on the region's agricultural productivity.

The Central Arizona Project (CAP) is a 336-mile (541 km) long canal that diverts 489 billion US gallons (1.85×10^9 m³) a year from the Colorado River to irrigate more than 300,000 acres (1,200 km²) of farmland. The CAP project also provides drinking water for Phoenix and Tucson. It has been estimated that Lake Mead, which dams the Colorado, has a 50-50 chance of running dry by 2021.

The Ipswich River near Boston now runs dry in some years due to heavy pumping of groundwater for irrigation. Maryland, Virginia and the District of Columbia have been fighting over the Potomac River. In drought years like 1999 or 2003, and on hot summer days the region consumes up to 85 percent of the river's flow.

Per Capita Withdrawal of Water

Turkmenistan, Kazakhstan and Uzbekistan use the most water per capita. See the table below:

Freshwater withdrawal by country and sector (top 15 countries, per capita)					
Country	Total freshwater withdrawal	Per capita withdrawal	Domestic use	Industrial use	Agricultural use
	(km³/yr)	(m³/p/yr)	(%)	(%)	(%)
Turkmenistan	24.65	5,104	2	1	98

Kazakhstan	35	2,360	2	17	82
Uzbekistan	58.34	2,194	5	2	93
Guyana	1.64	2,187	2	1	98
Hungary	21.03	2,082	9	59	32
Azerbaijan	17.25	2,051	5	28	68
Kyrgyzstan	10.08	1,916	3	3	94
Tajikistan	11.96	1,837	4	5	92
USA	477	1,600	13	46	41
Suriname	0.67	1,489	4	3	93
Iraq	42.7	1,482	3	5	92
Canada	44.72	1,386	20	69	12
Thailand	82.75	1,288	2	2	95
Ecuador	16.98	1,283	12	5	82
Australia	24.06	1,193	15	10	75

Turkmenistan

Orphaned ship in former Aral Sea, near Aral, Kazakhstan

Turkmenistan gets most of its water from the Amu Darya River. The Qaraqum Canal is a canal system that takes water from the Amu Darya River and distributes the water out over the desert for irrigation of its orchard crops and cotton. Turkmenistan uses the most water per capita in the world because only 55% of the water delivered to the fields actually reaches the crops.

Kazakhstan and Uzbekistan

The two rivers feeding the Aral Sea were dammed up and the water was diverted to irrigate the desert so that cotton could be produced. As a result, the Aral Sea's water has become much saltier and the sea's water level has decreased by over 60%. Drinking water is now contaminated with pesticides and other agricultural chemicals and contains bacteria and viruses. The climate has become more extreme in the area surrounding it.

Water Shortfall by Country

Saudi Arabia, Libya, Yemen and United Arab Emirates have hit peaks in water production and are depleting their water supply. See the table below:

Freshwater shortfall by country (top 15 countries)			
Region and country	**Total freshwater withdrawal** **(km³/yr)**	**Total freshwater supply** **(km³/yr)**	**Total freshwater shortfall** **(km³/yr)**
Saudi Arabia	17.32	2.4	14.9
Libya	4.27	0.6	3.7
Yemen	6.63	4.1	2.5
United Arab Emirates	2.3	0.2	2.2
Kuwait	0.44	0.02	0.4
Oman	1.36	1.0	0.4
Israel	2.05	1.7	0.4
Qatar	0.29	0.1	0.2
Bahrain	0.3	0.1	0.2
Jordan	1.01	0.9	0.1
Barbados	0.09	0.1	0.0
Maldives	0.003	0.03	0.0
Antigua and Barbuda	0.005	0.1	0.0
Malta	0.02	0.07	-0.1
Cyprus	0.21	0.4	-0.2

Saudi Arabia

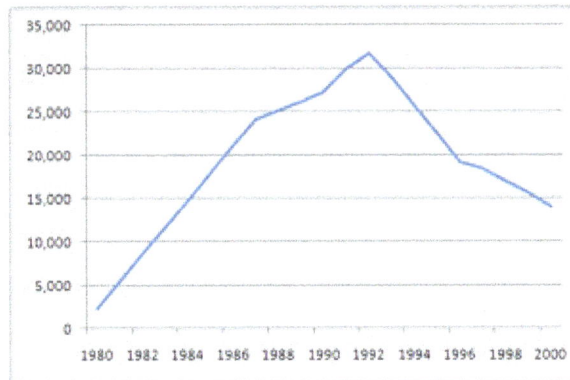

Water supply in Saudi Arabia, 1980–2000, in millions of cubic meters.

According to Walid A. Abderrahman (2001), "Water Demand Management in Saudi Arabia", Saudi Arabia reached peak water in the early 1990s, at more than 30 billion cubic meters per year, and declined afterward. The peak had arrived at about midpoint, as expected for a Hubbert curve.

Today, the water production is about half the peak rate. Saudi Arabian food production has been based on "fossil water"—water from ancient aquifers that is being recharged very slowly, if at all. Like oil, fossil water is non-renewable, and it is bound to run out someday. Saudi Arabia has abandoned its self-sufficient food production and is now importing virtually all of its food. Saudi Arabia has built desalination plants to provide about half the country's freshwater. The remainder comes from groundwater (40%), surface water (9%) and reclaimed wastewater (1%).

Libya

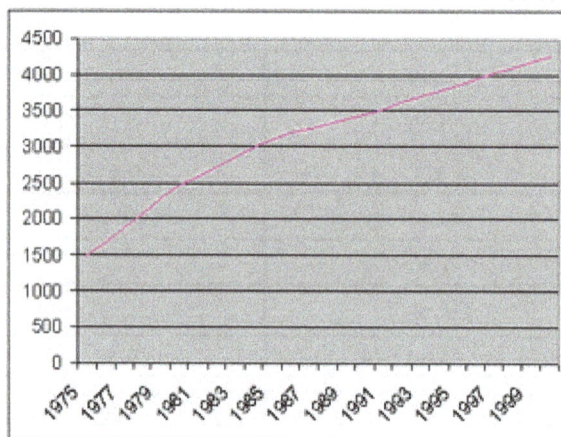

Water supply in Libya, 1975–2000, in millions of cubic meters.

Libya is working on a network of water pipelines to import water, called the Great Manmade River. It carries water from wells tapping fossil water in the Sahara desert to the cities of Tripoli, Benghazi, Sirte and others. Their water also comes from desalination plants.

Yemen

Peak water has occurred in Yemen. Sustainability is no longer attainable in Yemen, according to the government's five-year water plan for 2005–2009. The aquifer that supplies Sana'a, the capital of Yemen, will be depleted by 2009. In its search for water in the basin, the Yemeni government has drilled test wells that are 2 kilometers (1.2 mi) deep, depths normally associated with the oil industry, but it has failed to find water. Yemen must soon choose between relocating the city and building a pipeline to coastal desalination plants. The pipeline option is complicated by Sana'a's altitude of 2,250 m (7,380 ft).

As of 2010, the threat of running out of water was considered greater than that of Al-Qaeda or instability. There was speculation that Yemenis would have to abandon mountain cities, including Sana'a, and move to the coast. The cultivation of khat and poor water regulation by the government were partly blamed.

United Arab Emirates

United Arab Emirates has a rapidly growing economy and very little water to support it. UAE requires more water than is naturally available. They have reached peak water. To solve this, UAE has a desalination plant near Ruwais and ships its water via pipeline to Abu Dhabi.

Desalination plant in Ras al-Khaimah, UAE

Consequences

Famine

Water shortage may cause famine in Pakistan. Pakistan has approximately 35 million acres (140,000 km²) of arable land irrigated by canals and tube wells, mostly using water from the Indus River. Dams were constructed at Chashma, Mangla, and Tarbela to feed the irrigation system. Since the completion of the Tarbela Dam in 1976 no new capacity has been added despite astronomical growth in population. The gross capacity of the three dams has decreased because of sedimentation, a continual process. Per-capita surface-water availability for irrigation was 5,260 cubic meters per year in 1951. This has been reduced to a mere 1,100 cubic meters per year in 2006.

Health Problems

The quality of drinking water is vital for human health. Peak water constraints result in people not having access to safe water for basic personal hygiene. "Infectious waterborne diseases such as diarrhea, typhoid, and cholera are responsible for 80 percent of illnesses and deaths in the developing world, many of them children. One child dies every eight seconds from a waterborne disease; 15 million children a year."

Vital aquifers everywhere are becoming contaminated with toxins. Once an aquifer is contaminated, it is not likely that it can ever recover. Contaminants are more likely to cause chronic health effects. Water can be contaminated from pathogens such as bacteria, viruses, and parasites. Also, toxic organic chemicals can be a source of water contamination. Inorganic contaminants include toxic metals like arsenic, barium, chromium, lead, mercury, and silver. Nitrates are another source of inorganic contamination. Finally, leaching radioactive elements into the water supply can contaminate it.

Human Conflicts Over Water

Some conflicts of the future may be fought over the availability, quality, and control of water. Water has also been used as a tool in conflicts or as a target during conflicts that start for other reasons. Water shortages may well result in water conflicts over this precious resource.

In West Africa and other places like Nepal, Bangladesh, India (such as the Ganges Delta), and Peru, major changes in the rivers generate a significant risk of violent conflict in coming years. Water management and control could play a part in future resource wars over scarce resources.

Solutions

Freshwater usage has great potential for better conservation and management as it is used inefficiently nearly everywhere, but until actual scarcity hits, people tend to take access to freshwater for granted.

Water Conservation

There are several ways to reduce the use of water. For example, most irrigation systems waste water; typically, only between 35% and 50% of water withdrawn for irrigated agriculture ever reaches the crops. Most soaks into unlined canals, leaks out of pipes, or evaporates before reaching (or after being applied to) the fields. Swales and cisterns can be used to catch and store excess rainwater.

Water should be used more efficiently in industry, which should use a closed water cycle if possible. Also, industry should prevent polluting water so that it can be returned into the water cycle. Whenever possible, gray wastewater should be used to irrigate trees or lawns. Water drawn from aquifers should be recharged by treating the wastewater and returned to the aquifer.

Water can be conserved by not allowing freshwater to be used to irrigate luxuries such as golf courses. Luxury goods should not be produced in areas where freshwater has been depleted. For example, 1,500 liters of water is used on average for the manufacturing of a single computer and monitor.

Water Management

Sustainable water management involves the scientific planning, developing, distribution and optimization of water resources under defined water polices and regulations. Examples of policies that improve water management include the use of technology for efficiency monitoring and use of water, innovative water prices and markets, irrigation efficiency techniques, and much more.

Experience shows that higher water prices lead to improvements in the efficiency of use—a classical argument in economics, pricing, and markets. For example, Clark County, Nevada, raised its water rates in 2008 to encourage conservation. Economists propose to encourage conservation by adopting a system of progressive pricing whereby the price per unit of water used would start out very small, and then rise substantially for each additional unit of water used. This tiered-rate approach has been used for many years in many places, and is becoming more widespread. A Freakonomics column in the *New York Times* similarly suggested that people would respond to higher water prices by using less of it, just as they respond to higher gasoline prices by using less of it. The *Christian Science Monitor* has also reported on arguments that higher water prices curb waste and consumption.

Conversely, certain subsidies can lead to inefficient use of water. Water subsidies often involve contentious policy issues that are political in nature. In 2004, the Environmental Working Group criticized the United States government for selling subsidized water to corporate farms for an average price of only $17 per acre foot (326,000 gallons).

In his book *The Ultimate Resource 2*, Julian Simon claimed that there is a strong correlation between government corruption and lack of sufficient supplies of safe, clean water. Simon wrote, "there is complete agreement among water economists that all it takes to ensure an adequate supply for agriculture as well as for households in rich countries is that there be a rational structure of water law and market pricing. The problem is not too many people but rather defective laws and bureaucratic interventions; freeing up markets in water would eliminate just about all water problems forever... In poor water-short countries the problem with water supply—as with so many other matters—is lack of wealth to create systems to supply water efficiently enough. As these countries become richer, their water problems will become less difficult". This theoretical argument, however, ignores real-world conditions, including strong barriers to open water markets, the difficulty of moving water from one region to another, water rights laws that prevent redistribution based solely on economic value, inability of some populations to pay for water, and grossly imperfect information on water use. Actual experience with peak water constraints in some wealthy, but water-short countries and regions still suggests serious difficulties in reducing water challenges.

Climate Change

Extensive research has shown the direct links between water resources, the hydrologic cycle, and climatic change. As climate changes, there will be substantial impacts on water demands, precipitation patterns, storm frequency and intensity, snowfall and snowmelt dynamics, and more. Evidence from the IPCC to Working Group II, has shown climate change is already having a direct effect on animals, plants and water resources and systems. A 2007 report by the Intergovernmental Panel on Climate Change counted 75 million to 250 million people across Africa who could face water shortages by 2020. Crop yields could increase by 20% in East and Southeast Asia, but decrease by up to 30% in Central and South Asia. Agriculture fed by rainfall could drop by 50% in some African countries by 2020. A wide range of other impacts could affect peak water constraints.

Loss of biodiversity can be attributed largely to the appropriation of land for agroforestry and the effects of climate change. The 2008 IUCN Red List warns that long-term droughts and extreme weather puts additional stress on key habitats and, for example, lists 1,226 bird species as threatened with extinction, which is one-in-eight of all bird species.

Backstop Water Sources

The concept of a "backstop" resource is a resource that is sufficiently abundant and sustainable to replace non-renewable resources. Thus, solar and other renewable energy sources are considered "backstop" energy options for unsustainable fossil fuels. Similarly, Gleick and Palaniappan defined "backstop water sources" to be those resources that can replace unsustainable and non-renewable use of water, albeit typically at a higher cost. The classic backstop water source is desalination of seawater. If the rate of water production is not sufficient in one area, another "backstop" could be increased interbasin transfers, such as pipelines to carry freshwater from where it is abundant to an area where water is needed. Water can be imported into an area using water trucks. The most expensive and last resort measures of getting water to a community such as desalination, water transfers are called "backstop" water sources. Fog catchers are the most extreme of backstop methods.

To produce that fresh water, it can be obtained from ocean water through desalination. A January 17, 2008 article in the *Wall Street Journal* stated, "World-wide, 13,080 desalination plants produce more than 12 billion US gallons (45,000,000 m³) of water a day, according to the International Desalination Association". Israel is now desalinizing water at a cost of US$0.53 per cubic meter. Singapore is desalinizing water for US$0.49 per cubic meter. After being desalinized at Jubail, Saudi Arabia, water is pumped 200 miles (320 km) inland though a pipeline to the capital city of Riyadh.

However, several factors prevent desalination from being a panacea for water shortages:

- High capital costs to build the desalination plant

- High cost of the water produced

- Energy required to desalinate the water

- Environmental issues with the disposal of the resulting brine

- High cost of transporting water

Nevertheless, some countries like Spain are increasingly relying on desalination because of the continuing decreasing costs of the technology.

At last resort, it is possible in some particular regions to harvest water from fog using nets. The water from the nets drips into a tube. The tubes from several nets lead to a holding tank. Using this method, small communities on the edge of deserts can get water for drinking, gardening, showering and clothes washing. Critics say that fog catchers work in theory but have not succeeded as well in practice. This is due to the high expense of the nets and pipe, high maintenance costs and low quality of water.

An alternative approach is that of the Seawater Greenhouse, which consists in desalinating seawater through evaporation and condensation inside a greenhouse solely using solar energy. Successful pilots have been conducted growing crops in desert locations.

Saltwater Intrusion

Saltwater intrusion is the movement of saline water into freshwater aquifers, which can lead to contamination of drinking water sources and other consequences. Saltwater intrusion occurs naturally to some degree in most coastal aquifers, owing to the hydraulic connection between groundwater and seawater. Because saltwater has a higher mineral content than freshwater, it is denser and has a higher water pressure. As a result, saltwater can push inland beneath the freshwater. Certain human activities, especially groundwater pumping from coastal freshwater wells, have increased saltwater intrusion in many coastal areas. Water extraction drops the level of fresh groundwater, reducing its water pressure and allowing saltwater to flow further inland. Other contributors to saltwater intrusion include navigation channels or agricultural and drainage channels, which provide conduits for saltwater to move inland, and sea level rise. Saltwater intrusion can also be worsened by extreme events like hurricane storm surges.

Hydrology

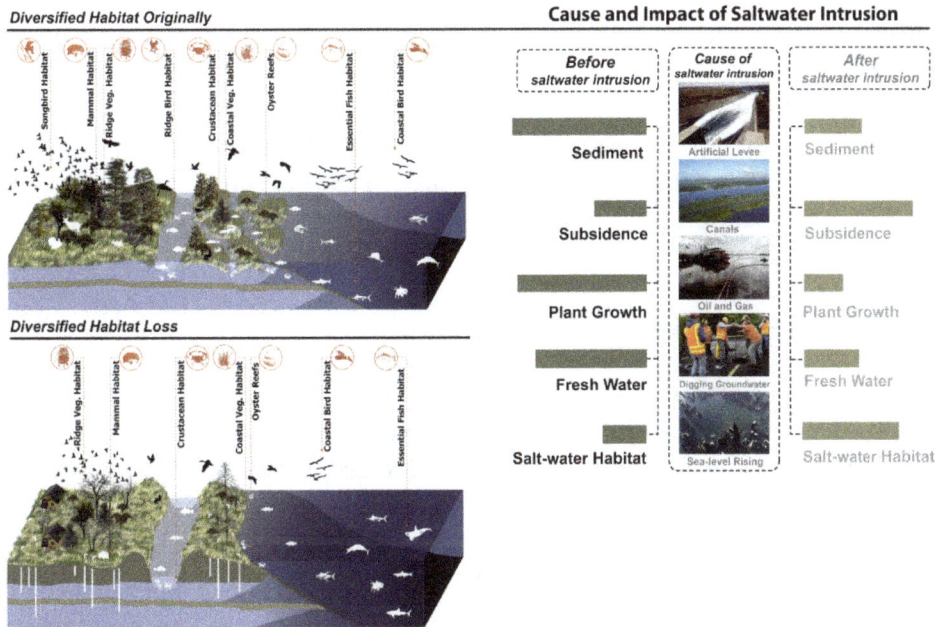

Cause and Impact of Saltwater Intrusion

At the coastal margin, fresh groundwater flowing from inland areas meets with saline groundwater from the ocean. The fresh groundwater flows from inland areas towards the coast where elevation and groundwater levels are lower. Because saltwater has a higher content of dissolved salts and minerals, it is denser than freshwater, causing it to have higher hydraulic head than freshwater. Hydraulic head refers to the liquid pressure exerted by a water column: a water column with higher hydraulic head will move into a water column with lower hydraulic head, if the columns are connected.

The higher pressure and density of saltwater causes it to move into coastal aquifers in a wedge shape under the freshwater. The saltwater and freshwater meet in a transition zone where mixing occurs through dispersion and diffusion. Ordinarily the inland extent of the saltwater wedge is limited because fresh groundwater levels, or the height of the freshwater column, increases as land elevation gets higher.

Causes

Groundwater Extraction

Groundwater extraction is the primary cause of saltwater intrusion. Groundwater is the main source of drinking water in many coastal areas of the United States, and extraction has increased over time. Under baseline conditions, the inland extent of saltwater is limited by higher pressure exerted by the freshwater column, owing to its higher elevation. Groundwater extraction can lower the level of the freshwater table, reducing the pressure exerted by the freshwater column and allowing the denser saltwater to move inland laterally. In Cape May, New Jersey, since the 1940s water withdrawals have lowered groundwater levels by up to 30 meters, reducing the water table to below sea level and causing widespread intrusion and contamination of water supply wells.

Groundwater extraction can also lead to well contamination by causing upwelling, or upcoming, of saltwater from the depths of the aquifer. Under baseline conditions, a saltwater wedge extends inland, underneath the freshwater because of its higher density. Water supply wells located over or near the saltwater wedge can draw the saltwater upward, creating a saltwater cone that might reach and contaminate the well. Some aquifers are predisposed towards this type of intrusion, such as the Lower Floridan aquifer: though a relatively impermeable rock or clay layer separates fresh groundwater from saltwater, isolated cracks breach the confining layer, promoting upward movement of saltwater. Pumping of groundwater strengthens this effect by lowering the water table, reducing the downward push of freshwater.

Canals and Drainage Networks

The construction of canals and drainage networks can lead to saltwater intrusion. Canals provide conduits for saltwater to be carried inland, as does the deepening of existing channels for navigation purposes. In Sabine Lake Estuary in the Gulf of Mexico, large-scale waterways have allowed saltwater to move into the lake, and upstream into the rivers feeding the lake. Additionally, channel dredging in the surrounding wetlands to facilitate oil and gas drilling has caused land subsidence, further promoting inland saltwater movement.

Drainage networks constructed to drain flat coastal areas can lead to intrusion by lowering the freshwater table, reducing the water pressure exerted by the freshwater column. Saltwater intrusion in southeast Florida has occurred largely as a result of drainage canals built between 1903 into the 1980s to drain the Everglades for agricultural and urban development. The main cause of intrusion was the lowering of the water table, though the canals also conveyed seawater inland until the construction of water control gates.

Effect on Water Supply

Many coastal communities around the United States are experiencing saltwater contamination of water supply wells, and this problem has been seen for decades. The consequences of saltwater intrusion for supply wells vary widely, depending on extent of the intrusion, the intended use of the water, and whether the salinity exceeds standards for the intended use. In some areas such as Washington State, intrusion only reaches portions of the aquifer, affecting only certain water supply wells. Other aquifers have faced more widespread salinity contamination, significantly affecting groundwater supplies for the region. For instance, in Cape May, New Jersey, where groundwater extraction has lowered water tables by up to 30 meters, saltwater intrusion has caused closure of over 120 water supply wells since the 1940s.

Ghyben–herzberg Relation

The first physical formulations of saltwater intrusion were made by W. Badon-Ghijben (1888, 1889) and A. Herzberg (1901), thus called the Ghyben–Herzberg relation. They derived analytical solutions to approximate the intrusion behavior, which are based on a number of assumptions that do not hold in all field cases.

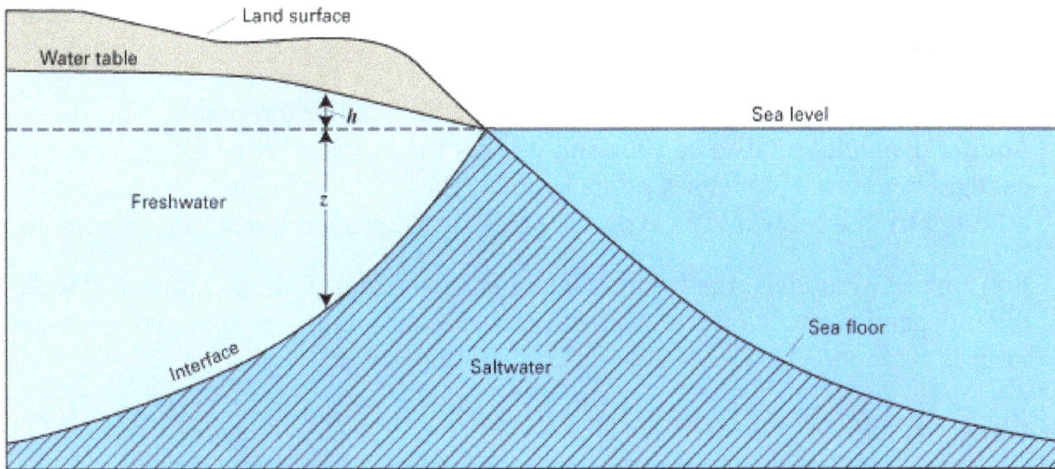

The figure shows the Ghyben–Herzberg relation. In the equation,

$$z = \frac{\rho_f}{(\rho_s - \rho_f)} h$$

the thickness of the freshwater zone above sea level is represented as and that below sea level is represented as . The two thicknesses and , are related by and where is the density of freshwater and is the density of saltwater. Freshwater has a density of about 1.000 grams per cubic centimeter (g/cm³) at 20 °C, whereas that of seawater is about 1.025 g/cm³. The equation can be simplified to

$$z = 40h$$

The Ghyben–Herzberg ratio states, for every foot of fresh water in an unconfined aquifer above sea level, there will be forty feet of fresh water in the aquifer below sea level.

In the 20th century the vastly increased computing power available allowed the use of numerical methods (usually finite differences or finite elements) that need fewer assumptions and can be applied more generally.

Modeling

Modeling of saltwater intrusion is considered difficult. Some typical difficulties that arise are:

- The possible presence of fissures and cracks and fractures in the aquifer, whose precise positions are unknown but which have great influence on the development of the saltwater intrusion

- The possible presence of small scale heterogeneities in the hydraulic properties of the aquifer, which are too small to be taken into account by the model but which may also have great influence on the development of the saltwater intrusion

- The change of hydraulic properties by the saltwater intrusion. A mixture of saltwater and freshwater is often undersaturated with respect to calcium, triggering dissolution of calcium in the mixing zone and changing hydraulic properties.

- The process known as cation exchange, which slows the advance of a saltwater intrusion and also slows the retreat of a saltwater intrusion.

- The fact that saltwater intrusions are often not in equilibrium makes it harder to model. Aquifer dynamics tend to be slow and it takes the intrusion cone a long time to adapt to changes in pumping schemes, rainfall, etc. So the situation in the field can be significantly different from what would be expected based on the sea level, pumping scheme etc.

- For long-term models, the future climate change forms a large unknown. Model results often depend strongly on sea level and recharge rate. Both are expected to change in the future.

Mitigation

Catfish Point control structure (lock) on the Mermentau River in coastal Louisiana

Saltwater is also an issue where a lock separates saltwater from freshwater (for example the Hiram M. Chittenden Locks in Washington). In this case a collection basin was built from which the saltwater can be pumped back to the sea. Some of the intruding saltwater is also pumped to the fish ladder to make it more attractive to migrating fish.

Areas where saltwater intrusion is occurring

- Water supply and sanitation in Benin

- Geography of Cyprus

- Bou Regreg (Morocco)

- Water resources management in Pakistan

- Water supply and sanitation in Tunisia

- United States

o   ACF River Basin (Florida/Georgia)

o   Environment of Florida

o   Hiram M. Chittenden Locks (Washington)

o   Hutchinson Island (Georgia)

o   Lake Lanier (Georgia)

o   Lake Pontchartrain (Louisiana)

o   Miami River (Florida)

o   Oxnard Plain (California)

o   San Leandro (California)

o   Sonoma Creek (California)

Wastewater

Wastewater treatment plant in Cuxhaven, Germany

Wastewater, also written as waste water, is any water that has been adversely affected in quality by anthropogenic influence. Wastewater can originate from a combination of domestic, industrial, commercial or agricultural activities, surface runoff or stormwater, and from sewer inflow or infiltration.

Municipal wastewater (also called sewage) is usually conveyed in a combined sewer or sanitary sewer, and treated at a wastewater treatment plant. Treated wastewater is discharged into receiving water via an effluent pipe. Wastewaters generated in areas without access to centralized sewer systems rely on on-site wastewater systems. These typically comprise a septic tank, drain field, and optionally an on-site treatment unit. The management of wastewater belongs to the overarching term sanitation, just like the management of human excreta, solid waste and stormwater (drainage).

Sewage is a type of wastewater that comprises domestic wastewater and is therefore contaminated with feces or urine from people's toilets, but the term sewage is also used to mean any type of wastewater. Sewerage is the physical infrastructure, including pipes, pumps, screens, channels etc. used to convey sewage from its origin to the point of eventual treatment or disposal.

Origin

Wastewater can come from:

- Human excreta (feces and urine) often mixed with used toilet paper or wipes; this is known as blackwater if it is collected with flush toilets

- Washing water (personal, clothes, floors, dishes, cars, etc.), also known as greywater or sullage

- Surplus manufactured liquids from domestic sources (drinks, cooking oil, pesticides, lubricating oil, paint, cleaning liquids, etc.)

- Urban rainfall runoff from roads, carparks, roofs, sidewalks/pavements (contains oils, animal feces, litter, gasoline/petrol, diesel or rubber residues from tires, soapscum, metals from vehicle exhausts, etc.)

- Highway drainage (oil, de-icing agents, rubber residues, particularly from tires)

- Storm drains (may include trash)

- Manmade liquids (illegal disposal of pesticides, used oils, etc.)

- Industrial waste

- Industrial site drainage (silt, sand, alkali, oil, chemical residues);

  o Industrial cooling waters (biocides, heat, slimes, silt)

  o Industrial process waters

  o Organic or biodegradable waste, including waste from abattoirs, creameries, and ice cream manufacture

  o Organic or non bio-degradable/difficult-to-treat waste (pharmaceutical or pesticide manufacturing)

  o Extreme pH waste (from acid/alkali manufacturing, metal plating)

  o Toxic waste (metal plating, cyanide production, pesticide manufacturing, etc.)

  o Solids and emulsions (paper manufacturing, foodstuffs, lubricating and hydraulic oil manufacturing, etc.)

- o   Agricultural drainage, direct and diffuse
- o   Hydraulic fracturing
- o   Produced water from oil & natural gas production

Wastewater can be diluted or mixed with other types of water in the form of:

- Seawater ingress (high volumes of salt and microbes)
- Direct ingress of river water
- Rainfall collected on roofs, yards, hard-standings, etc. (generally clean with traces of oils and fuel)
- Groundwater infiltrated into sewage

After it has undergone some treatment, the "treated wastewater" remains, e.g.:

- Septic tank discharge
- Sewage treatment plant discharge

Constituents

The composition of wastewater varies widely. This is a partial list of what it may contain:

- Water (more than 95 percent), which is often added during flushing to carry waste down a drain;
- Pathogens such as bacteria, viruses, prions and parasitic worms;
- Non-pathogenic bacteria;
- Organic particles such as feces, hairs, food, vomit, paper fibers, plant material, humus, etc.;
- Soluble organic material such as urea, fruit sugars, soluble proteins, drugs, pharmaceuticals, etc.;
- Inorganic particles such as sand, grit, metal particles, ceramics, etc.;
- Soluble inorganic material such as ammonia, road-salt, sea-salt, cyanide, hydrogen sulfide, thiocyanates, thiosulfates, etc.;
- Animals such as protozoa, insects, arthropods, small fish, etc.;
- Macro-solids such as sanitary napkins, nappies/diapers, condoms, needles, children's toys, dead animals or plants, etc.;
- Gases such as hydrogen sulfide, carbon dioxide, methane, etc.;
- Emulsions such as paints, adhesives, mayonnaise, hair colorants, emulsified oils, etc.;
- Toxins such as pesticides, poisons, herbicides, etc.

- Pharmaceuticals and hormones and other hazardous substances

Quality Indicators

Any oxidizable material present in an aerobic natural waterway or in an industrial wastewater will be oxidized both by biochemical (bacterial) or chemical processes. The result is that the oxygen content of the water will be decreased. Basically, the reaction for biochemical oxidation may be written as:

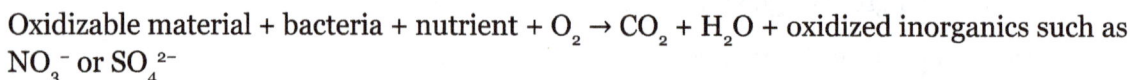

Oxidizable material + bacteria + nutrient + $O_2 \rightarrow CO_2 + H_2O$ + oxidized inorganics such as NO_3^- or SO_4^{2-}

Oxygen consumption by reducing chemicals such as sulfides and nitrites is typified as follows:

$$S^{2-} + 2\,O_2 \rightarrow SO_4^{2-}$$

$$NO_2^- + \tfrac{1}{2}\,O_2 \rightarrow NO_3^-$$

Since all natural waterways contain bacteria and nutrients, almost any waste compounds introduced into such waterways will initiate biochemical reactions (such as shown above). Those biochemical reactions create what is measured in the laboratory as the biochemical oxygen demand (BOD). Such chemicals are also liable to be broken down using strong oxidizing agents and these chemical reactions create what is measured in the laboratory as the chemical oxygen demand (COD). Both the BOD and COD tests are a measure of the relative oxygen-depletion effect of a waste contaminant. Both have been widely adopted as a measure of pollution effect. The BOD test measures the oxygen demand of biodegradable pollutants whereas the COD test measures the oxygen demand of oxidizable pollutants.

The so-called 5-day BOD measures the amount of oxygen consumed by biochemical oxidation of waste contaminants in a 5-day period. The total amount of oxygen consumed when the biochemical reaction is allowed to proceed to completion is called the Ultimate BOD. Because the Ultimate BOD is so time consuming, the 5-day BOD has been almost universally adopted as a measure of relative pollution effect.

There are also many different COD tests of which the 4-hour COD is probably the most common.

There is no generalized correlation between the 5-day BOD and the ultimate BOD. Similarly there is no generalized correlation between BOD and COD. It is possible to develop such correlations for specific waste contaminants in a specific wastewater stream but such correlations cannot be generalized for use with any other waste contaminants or wastewater streams. This is because the composition of any wastewater stream is different. As an example an effluent consisting of a solution of simple sugars that might discharge from a confectionery factory is likely to have organic components that degrade very quickly. In such a case, the 5 day BOD and the ultimate BOD would be very similar since there would be very little organic material left after 5 days. However a final effluent of a sewage treatment works serving a large industrialised area might have a discharge where the ultimate BOD was much greater than the 5 day BOD because much of the easily degraded material would have been removed in the sewage treatment process and many industrial processes discharge difficult to degrade organic molecules.

The laboratory test procedures for the determining the above oxygen demands are detailed in many standard texts. American versions include the "Standard Methods for the Examination of Water and Wastewater."

Disposal

Industrial wastewater effluent with neutralized pH from tailing runoff. Taken in Peru.

In some urban areas, municipal wastewater is carried separately in sanitary sewers and runoff from streets is carried in storm drains. Access to either of these is typically through a manhole. During high precipitation periods a combined sewer overflow can occur, forcing untreated sewage to flow back into the environment. This can pose a serious threat to public health and the surrounding environment.

Sewage may drain directly into major watersheds with minimal or no treatment but this usually has serious impacts on the quality of an environment and on the health of people. Pathogens can cause a variety of illnesses. Some chemicals pose risks even at very low concentrations and can remain a threat for long periods of time because of bioaccumulation in animal or human tissue.

Treatment

There are numerous processes that can be used to clean up wastewaters depending on the type and extent of contamination. Wastewater can be treated in wastewater treatment plants which include physical, chemical and biological treatment processes. Municipal wastewater is treated in sewage treatment plants (which may also be referred to as wastewater treatment plants). Agricultural wastewater may be treated in agricultural wastewater treatment processes, whereas industrial wastewater is treated in industrial wastewater treatment processes.

For municipal wastewater the use of septic tanks and other On-Site Sewage Facilities (OSSF) is widespread in some rural areas, for example serving up to 20 percent of the homes in the U.S.

One type of aerobic treatment system is the activated sludge process, based on the maintenance and recirculation of a complex biomass composed of micro-organisms able to absorb and adsorb the organic matter carried in the wastewater. Anaerobic wastewater treatment processes (UASB, EGSB) are also widely applied in the treatment of industrial wastewaters and biological sludge. Some wastewater may be highly treated and reused as reclaimed water. Constructed wetlands are also being used.

Reuse

Treated wastewater can be reused as drinking water, in industry (for example in cooling towers), in artificial recharge of aquifers, in agriculture and in the rehabilitation of natural ecosystems (for example in Florida's Everglades).

Reuse in Gardening and Agriculture

There are benefits of using recycled water for irrigation, including the lower cost compared to some other sources and consistency of supply regardless of season, climatic conditions and associated water restrictions. Irrigation with recycled wastewater can also serve to fertilize plants if it contains nutrients, such as nitrogen, phosphorus and potassium.

Around 90% of wastewater produced globally remains untreated, causing widespread water pollution, especially in low-income countries. Increasingly, agriculture is using untreated wastewater for irrigation. Cities provide lucrative markets for fresh produce, so are attractive to farmers. However, because agriculture has to compete for increasingly scarce water resources with industry and municipal users, there is often no alternative for farmers but to use water polluted with urban waste directly to water their crops.

Health Risks of Polluted Irrigation Water

There can be significant health hazards related to using untreated wastewater in agriculture. Wastewater from cities can contain a mixture of chemical and biological pollutants. In low-income countries, there are often high levels of pathogens from excreta, while in emerging nations, where industrial development is outpacing environmental regulation, there are increasing risks from inorganic and organic chemicals. The World Health Organization, in collaboration with the Food and Agriculture Organization of the United Nations (FAO) and the United Nations Environmental Program (UNEP), has developed guidelines for safe use of wastewater in 2006. These guidelines advocate a 'multiple-barrier' approach to wastewater use, for example by encouraging farmers to adopt various risk-reducing behaviours. These include ceasing irrigation a few days before harvesting to allow pathogens to die off in the sunlight, applying water carefully so it does not contaminate leaves likely to be eaten raw, cleaning vegetables with disinfectant or allowing fecal sludge used in farming to dry before being used as a human manure.

Etymology

The words "sewage" and "sewer" came from Old French *essouier* = "to drain", which came from Latin *exaquāre*. Their formal Latin antecedents are *exaquāticum* and *exaquārium*.

Legislation

European Union

Council Directive 91/271/EEC on Urban Wastewater Treatment was adopted on 21 May 1991, amended by the Commission Directive 98/15/EC. Commission Decision 93/481/EEC defines the information that Member States should provide the Commission on the state of implementation of the Directive.

United States

The Clean Water Act is the primary federal law in the United States governing water pollution.

References

- Van Ginkel, J. A. (2002). Human Development and the Environment: Challenges for the United Nations in the New Millennium. United Nations University Press. pp. 198–199. ISBN 9280810693.

- Prokurat, Sergiusz (2015), Drought and water shortages in Asia as a threat and economic problem (PDF), Józefów: "Journal of Modern Science" 3/26/2015, pp. 235–250, retrieved 5 August 2016

- IWMI Research Report 83. "Spatial variation in water supply and demand across river basins of India" (PDF). Retrieved 23 June 2015.

- Brown, Lester R. (November 29, 2013). "India's dangerous 'food bubble'". Los Angeles Times. Archived from the original on December 4, 2013. Retrieved July 13, 2014.

- David Grey and Claudia W. Sadoff - (2007-09-01). "Sink or Swim? Water security for growth and development". Water Policy. Iwaponline.com. 9 (6): 545–571. Retrieved 2014-08-16.

- Good, B. J., Buchtel, J., Meffert, D.J., Radford, J., Rhinehart, W., Wilson, R. (1995). "Louisiana's Major Coastal Navigation Channels" (pdf). Louisiana Department of Natural Resources. Retrieved 2013-09-14.

- Lacombe, Pierre J. & Carleton, Glen B. (2002). "Hydrogeologic Framework, Availability of Water Supplies, and Saltwater Intrusion, Cape May County, New Jersey" (PDF). USGS. Retrieved 2012-12-10.

- The World Bank, 2009 "Water and Climate Change: Understanding the Risks and Making Climate-Smart Investment Decisions". pp. 21–24. Retrieved 24 October 2011.

4

Significant Aspects of Water Pollution

The contamination of water bodies is water pollution. The aspects of water pollution dealt within this chapter are marine pollution, ocean acidification, thermal pollution, nutrient pollution and non point source pollution. The topics discussed in the chapter are of great importance to broaden the existing knowledge on water pollution.

Water Pollution

Water pollution is the contamination of water bodies (e.g. lakes, rivers, oceans, aquifers and groundwater). This form of environmental degradation occurs when pollutants are directly or indirectly discharged into water bodies without adequate treatment to remove harmful compounds.

Raw sewage and industrial waste in the New River as it passes from Mexicali to Calexico, California

Water pollution affects the entire biosphere – plants and organisms living in these bodies of water. In almost all cases the effect is damaging not only to individual species and population, but also to the natural biological communities.

Introduction

Water pollution is a major global problem which requires ongoing evaluation and revision of water resource policy at all levels (international down to individual aquifers and wells). It has been suggested that water pollution is the leading worldwide cause of deaths and diseases, and that it accounts for the deaths of more than 14,000 people daily. An estimated 580 people in India die of water pollution related illness every day. About 90 percent of the water in the cities of China is polluted. As of 2007, half a billion Chinese had no access to safe drinking water. In addition to the acute problems of water pollution in developing countries, developed countries also continue to struggle with pollution problems. For example, in the most recent national report on water quality in the United States, 44 percent of assessed stream miles, 64 percent of assessed lake acres, and 30 percent of assessed bays and estuarine square miles were classified as polluted. The head of China's national development agency said in 2007 that one quarter the length of China's seven main rivers were so poisoned the water harmed the skin.

Pollution in the Lachine Canal, Canada

Water is typically referred to as polluted when it is impaired by anthropogenic contaminants and either does not support a human use, such as drinking water, or undergoes a marked shift in its ability to support its constituent biotic communities, such as fish. Natural phenomena such as volcanoes, algae blooms, storms, and earthquakes also cause major changes in water quality and the ecological status of water.

Categories

Although interrelated, surface water and groundwater have often been studied and managed as separate resources. Surface water seeps through the soil and becomes groundwater. Conversely, groundwater can also feed surface water sources. Sources of surface water pollution are generally grouped into two categories based on their origin.

Point Sources

Point source water pollution refers to contaminants that enter a waterway from a single, identifiable source, such as a pipe or ditch. Examples of sources in this category include discharges from a sewage treatment plant, a factory, or a city storm drain. The U.S. Clean Water Act (CWA) defines point source for regulatory enforcement purposes. The CWA definition of point source was amended in 1987 to include municipal storm sewer systems, as well as industrial storm water, such as from construction sites.

Point source pollution – Shipyard – Rio de Janeiro.

Non-point Sources

Nonpoint source pollution refers to diffuse contamination that does not originate from a single discrete source. NPS pollution is often the cumulative effect of small amounts of contaminants gathered from a large area. A common example is the leaching out of nitrogen compounds from fertilized agricultural lands. Nutrient runoff in storm water from "sheet flow" over an agricultural field or a forest are also cited as examples of NPS pollution.

Blue drain and yellow fish symbol used by the UK Environment Agency to raise awareness of the ecological impacts of contaminating surface drainage

Contaminated storm water washed off of parking lots, roads and highways, called urban runoff, is sometimes included under the category of NPS pollution. However, because this runoff is typically channeled into storm drain systems and discharged through pipes to local surface waters, it becomes a point source.

Groundwater Pollution

Interactions between groundwater and surface water are complex. Consequently, groundwater pollution, also referred to as groundwater contamination, is not as easily classified as surface water pollution. By its very nature, groundwater aquifers are susceptible to contamination from sources that may not directly affect surface water bodies, and the distinction of point vs. non-point source may be irrelevant. A spill or ongoing release of chemical or radionuclide contaminants into soil (located away from a surface water body) may not create point or non-point source pollution but can contaminate the aquifer below, creating a toxic plume. The movement of the plume, called a plume front, may be analyzed through a hydrological transport model or groundwater model. Analysis of groundwater contamination may focus on soil characteristics and site geology, hydrogeology, hydrology, and the nature of the contaminants.

Causes

The specific contaminants leading to pollution in water include a wide spectrum of chemicals, pathogens, and physical changes such as elevated temperature and discoloration. While many of the chemicals and substances that are regulated may be naturally occurring (calcium, sodium, iron, manganese, etc.) the concentration is often the key in determining what is a natural component of water and what is a contaminant. High concentrations of naturally occurring substances can have negative impacts on aquatic flora and fauna.

Oxygen-depleting substances may be natural materials such as plant matter (e.g. leaves and grass) as well as man-made chemicals. Other natural and anthropogenic substances may cause turbidity (cloudiness) which blocks light and disrupts plant growth, and clogs the gills of some fish species.

Many of the chemical substances are toxic. Pathogens can produce waterborne diseases in either human or animal hosts. Alteration of water's physical chemistry includes acidity (change in pH), electrical conductivity, temperature, and eutrophication. Eutrophication is an increase in the concentration of chemical nutrients in an ecosystem to an extent that increases in the primary productivity of the ecosystem. Depending on the degree of eutrophication, subsequent negative environmental effects such as anoxia (oxygen depletion) and severe reductions in water quality may occur, affecting fish and other animal populations.

Pathogens

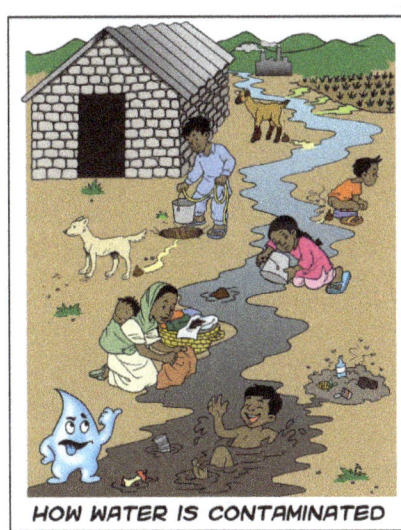

HOW WATER IS CONTAMINATED

Poster to teach people in South Asia about human activities leading to the pollution of water sources

A manhole cover unable to contain a sanitary sewer overflow.

Fecal sludge collected from pit latrines is dumped into a river at the Korogocho slum in Nairobi, Kenya.

Disease-causing microorganisms are referred to as pathogens. Although the vast majority of bacteria are either harmless or beneficial, a few pathogenic bacteria can cause disease. Coliform bacteria, which are not an actual cause of disease, are commonly used as a bacterial indicator of water pollution. Other microorganisms sometimes found in surface waters that have caused human health problems include:

- *Burkholderia pseudomallei*

- *Cryptosporidium parvum*

- *Giardia lamblia*

- *Salmonella*

- *Norovirus* and other viruses

- *Parasitic worms including the Schistosoma type*

High levels of pathogens may result from on-site sanitation systems (septic tanks, pit latrines) or inadequately treated sewage discharges. This can be caused by a sewage plant designed with less than secondary treatment (more typical in less-developed countries). In developed countries, older cities with aging infrastructure may have leaky sewage collection systems (pipes, pumps, valves), which can cause sanitary sewer overflows. Some cities also have combined sewers, which may discharge untreated sewage during rain storms.

Pathogen discharges may also be caused by poorly managed livestock operations.

Organic, Inorganic and Macroscopic Contaminants

Contaminants may include organic and inorganic substances.

A garbage collection boom in an urban-area stream in Auckland, New Zealand.

Organic water pollutants include:

- Detergents

- Disinfection by-products found in chemically disinfected drinking water, such as chloroform

- Food processing waste, which can include oxygen-demanding substances, fats and grease

- Insecticides and herbicides, a huge range of organohalides and other chemical compounds

- Petroleum hydrocarbons, including fuels (gasoline, diesel fuel, jet fuels, and fuel oil) and lubricants (motor oil), and fuel combustion byproducts, from storm water runoff

- Volatile organic compounds, such as industrial solvents, from improper storage.

- Chlorinated solvents, which are dense non-aqueous phase liquids, may fall to the bottom of reservoirs, since they don't mix well with water and are denser.

    o Polychlorinated biphenyl (PCBs)

    o Trichloroethylene

- Perchlorate

- Various chemical compounds found in personal hygiene and cosmetic products

- Drug pollution involving pharmaceutical drugs and their metabolites

Inorganic water pollutants include:

- Acidity caused by industrial discharges (especially sulfur dioxide from power plants)

- Ammonia from food processing waste

- Chemical waste as industrial by-products

- Fertilizers containing nutrients--nitrates and phosphates—which are found in storm water runoff from agriculture, as well as commercial and residential use

- Heavy metals from motor vehicles (via urban storm water runoff) and acid mine drainage

- Silt (sediment) in runoff from construction sites, logging, slash and burn practices or land clearing sites.

Macroscopic pollution – large visible items polluting the water – may be termed "floatables" in an urban storm water context, or marine debris when found on the open seas, and can include such items as:

- Trash or garbage (e.g. paper, plastic, or food waste) discarded by people on the ground, along with accidental or intentional dumping of rubbish, that are washed by rainfall into storm drains and eventually discharged into surface waters

- Nurdles, small ubiquitous waterborne plastic pellets

- Shipwrecks, large derelict ships.

The Brayton Point Power Station in Massachusetts discharges heated water to Mount Hope Bay.

Thermal Pollution

Thermal pollution is the rise or fall in the temperature of a natural body of water caused by human influence. Thermal pollution, unlike chemical pollution, results in a change in the physical properties of water. A common cause of thermal pollution is the use of water as a coolant by power plants and industrial manufacturers. Elevated water temperatures decrease oxygen levels, which can kill fish and alter food chain composition, reduce species biodiversity, and foster invasion by new thermophilic species. Urban runoff may also elevate temperature in surface waters.

Thermal pollution can also be caused by the release of very cold water from the base of reservoirs into warmer rivers.

Transport and Chemical Reactions of Water Pollutants

Most water pollutants are eventually carried by rivers into the oceans. In some areas of the world the influence can be traced one hundred miles from the mouth by studies using hydrology transport models. Advanced computer models such as SWMM or the DSSAM Model have been used in many locations worldwide to examine the fate of pollutants in aquatic systems. Indicator filter feeding species such as copepods have also been used to study pollutant fates in the New York Bight, for example. The highest toxin loads are not directly at the mouth of the Hudson River, but 100 km (62 mi) south, since several days are required for incorporation into planktonic tissue. The Hudson discharge flows south along the coast due to the coriolis force. Further south are areas of oxygen depletion caused by chemicals using up oxygen and by algae blooms, caused by excess nutrients from algal cell death and decomposition. Fish and shellfish kills have been reported, because toxins climb the food chain after small fish consume copepods, then large fish eat smaller fish, etc. Each successive step up the food chain causes a cumulative concentration of pollutants such as heavy metals (e.g. mercury) and persistent organic pollutants such as DDT. This is known as bio-magnification, which is occasionally used interchangeably with bio-accumulation.

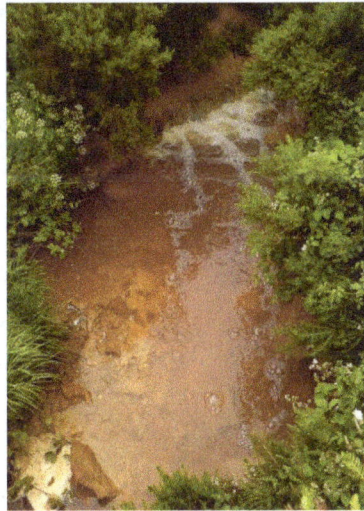

A polluted river draining an abandoned copper mine on Anglesey

Large gyres (vortexes) in the oceans trap floating plastic debris. The North Pacific Gyre, for example, has collected the so-called "Great Pacific Garbage Patch", which is now estimated to be one hundred times the size of Texas. Plastic debris can absorb toxic chemicals from ocean pollution, potentially poisoning any creature that eats it. Many of these long-lasting pieces wind up in the stomachs of marine birds and animals. This results in obstruction of digestive pathways, which leads to reduced appetite or even starvation.

Many chemicals undergo reactive decay or chemical change, especially over long periods of time in groundwater reservoirs. A noteworthy class of such chemicals is the chlorinated hydrocarbons such as trichloroethylene (used in industrial metal degreasing and electronics manufacturing) and tetrachloroethylene used in the dry cleaning industry. Both of these chemicals, which are carcinogens themselves, undergo partial decomposition reactions, leading to new hazardous chemicals (including dichloroethylene and vinyl chloride).

Groundwater pollution is much more difficult to abate than surface pollution because groundwater can move great distances through unseen aquifers. Non-porous aquifers such as clays partially purify water of bacteria by simple filtration (adsorption and absorption), dilution, and, in some cases, chemical reactions and biological activity; however, in some cases, the pollutants merely transform to soil contaminants. Groundwater that moves through open fractures and caverns is not filtered and can be transported as easily as surface water. In fact, this can be aggravated by the human tendency to use natural sinkholes as dumps in areas of karst topography.

There are a variety of secondary effects stemming not from the original pollutant, but a derivative condition. An example is silt-bearing surface runoff, which can inhibit the penetration of sunlight through the water column, hampering photosynthesis in aquatic plants.

Measurement

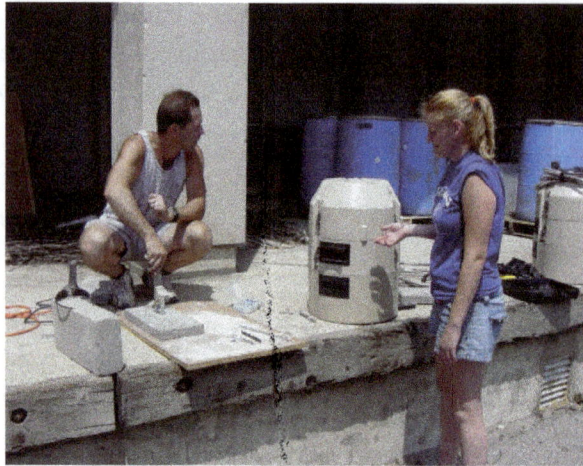

Environmental scientists preparing water autosamplers.

Water pollution may be analyzed through several broad categories of methods: physical, chemical and biological. Most involve collection of samples, followed by specialized analytical tests. Some methods may be conducted *in situ*, without sampling, such as temperature. Government agencies and research organizations have published standardized, validated analytical test methods to facilitate the comparability of results from disparate testing events.

Sampling

Sampling of water for physical or chemical testing can be done by several methods, depending on the accuracy needed and the characteristics of the contaminant. Many contamination events are sharply restricted in time, most commonly in association with rain events. For this reason "grab" samples are often inadequate for fully quantifying contaminant levels. Scientists gathering this type of data often employ auto-sampler devices that pump increments of water at either time or discharge intervals.

Sampling for biological testing involves collection of plants and/or animals from the surface water body. Depending on the type of assessment, the organisms may be identified for biosurveys (population counts) and returned to the water body, or they may be dissected for bioassays to determine toxicity.

Physical Testing

Common physical tests of water include temperature, solids concentrations (e.g., total suspended solids (TSS)) and turbidity.

Chemical Testing

Water samples may be examined using the principles of analytical chemistry. Many published test methods are available for both organic and inorganic compounds. Frequently used methods include pH, biochemical oxygen demand (BOD), chemical oxygen demand (COD), nutrients (nitrate and phosphorus compounds), metals (including copper, zinc, cadmium, lead and mercury), oil and grease, total petroleum hydrocarbons (TPH), and pesticides.

Biological Testing

Biological testing involves the use of plant, animal, and/or microbial indicators to monitor the health of an aquatic ecosystem. They are any biological species or group of species whose function, population, or status can reveal what degree of ecosystem or environmental integrity is present. One example of a group of bio-indicators are the copepods and other small water crustaceans that are present in many water bodies. Such organisms can be monitored for changes (biochemical, physiological, or behavioral) that may indicate a problem within their ecosystem.

Control of Pollution

Decisions on the type and degree of treatment and control of wastes, and the disposal and use of adequately treated wastewater, must be based on a consideration all the technical factors of each drainage basin, in order to prevent any further contamination or harm to the environment.

Sewage Treatment

Deer Island Wastewater Treatment Plant serving Boston, Massachusetts and vicinity.

In urban areas of developed countries, domestic sewage is typically treated by centralized sewage treatment plants. Well-designed and operated systems (i.e., secondary treatment or better) can remove 90 percent or more of the pollutant load in sewage. Some plants have additional systems to remove nutrients and pathogens.

Cities with sanitary sewer overflows or combined sewer overflows employ one or more engineering approaches to reduce discharges of untreated sewage, including:

- utilizing a green infrastructure approach to improve storm water management capacity throughout the system, and reduce the hydraulic overloading of the treatment plant

- repair and replacement of leaking and malfunctioning equipment

- increasing overall hydraulic capacity of the sewage collection system (often a very expensive option).

A household or business not served by a municipal treatment plant may have an individual septic tank, which pre-treats the wastewater on site and infiltrates it into the soil.

Industrial Wastewater Treatment

Dissolved air flotation system for treating industrial wastewater.

Some industrial facilities generate ordinary domestic sewage that can be treated by municipal facilities. Industries that generate wastewater with high concentrations of conventional pollutants (e.g. oil and grease), toxic pollutants (e.g. heavy metals, volatile organic compounds) or other non-conventional pollutants such as ammonia, need specialized treatment systems. Some of these facilities can install a pre-treatment system to remove the toxic components, and then send the partially treated wastewater to the municipal system. Industries generating large volumes of wastewater typically operate their own complete on-site treatment systems. Some industries have been successful at redesigning their manufacturing processes to reduce or eliminate pollutants, through a process called pollution prevention.

Heated water generated by power plants or manufacturing plants may be controlled with:

- cooling ponds, man-made bodies of water designed for cooling by evaporation, convection, and radiation

- cooling towers, which transfer waste heat to the atmosphere through evaporation and/or heat transfer

- cogeneration, a process where waste heat is recycled for domestic and/or industrial heating purposes.

Riparian buffer lining a creek in Iowa.

Agricultural Wastewater Treatment

Non Point Source Controls

Sediment (loose soil) washed off fields is the largest source of agricultural pollution in the United States. Farmers may utilize erosion controls to reduce runoff flows and retain soil on their fields. Common techniques include contour plowing, crop mulching, crop rotation, planting perennial crops and installing riparian buffers.

Nutrients (nitrogen and phosphorus) are typically applied to farmland as commercial fertilizer, animal manure, or spraying of municipal or industrial wastewater (effluent) or sludge. Nutrients may also enter runoff from crop residues, irrigation water, wildlife, and atmospheric deposition. Farmers can develop and implement nutrient management plans to reduce excess application of nutrients and reduce the potential for nutrient pollution.

To minimize pesticide impacts, farmers may use Integrated Pest Management (IPM) techniques (which can include biological pest control) to maintain control over pests, reduce reliance on chemical pesticides, and protect water quality.

Feedlot in the United States

Point Source Wastewater Treatment

Farms with large livestock and poultry operations, such as factory farms, are called *concentrated animal feeding operations* or *feedlots* in the US and are being subject to increasing government regulation. Animal slurries are usually treated by containment in anaerobic lagoons before disposal by spray or trickle application to grassland. Constructed wetlands are sometimes used to facilitate treatment of animal wastes. Some animal slurries are treated by mixing with straw and composted at high temperature to produce a bacteriologically sterile and friable manure for soil improvement.

Erosion and Sediment Control from Construction Sites

Silt fence installed on a construction site.

Sediment from construction sites is managed by installation of:

- erosion controls, such as mulching and hydroseeding, and

- sediment controls, such as sediment basins and silt fences.

Discharge of toxic chemicals such as motor fuels and concrete washout is prevented by use of:

- spill prevention and control plans, and

- specially designed containers (e.g. for concrete washout) and structures such as overflow controls and diversion berms.

Control of Urban Runoff (Storm Water)

Retention basin for controlling urban runoff

Effective control of urban runoff involves reducing the velocity and flow of storm water, as well as reducing pollutant discharges. Local governments use a variety of storm water management techniques to reduce the effects of urban runoff. These techniques, called best management practices (BMPs) in the U.S., may focus on water quantity control, while others focus on improving water quality, and some perform both functions.

Pollution prevention practices include low-impact development techniques, installation of green roofs and improved chemical handling (e.g. management of motor fuels & oil, fertilizers and pesticides). Runoff mitigation systems include infiltration basins, bioretention systems, constructed wetlands, retention basins and similar devices.

Thermal pollution from runoff can be controlled by storm water management facilities that absorb the runoff or direct it into groundwater, such as bioretention systems and infiltration basins. Retention basins tend to be less effective at reducing temperature, as the water may be heated by the sun before being discharged to a receiving stream.

Water Pollution by Country

Water pollution by country
• Albania
• Australia
• Brazil
• Canada
• China
• Egypt
• India
• Indonesia
• Iran
• Japan
• New Zealand
• Peru
• Philippines
• Russia
• Thailand
• Turkey
• United States
• Venezuela
• Vietnam

Marine Pollution

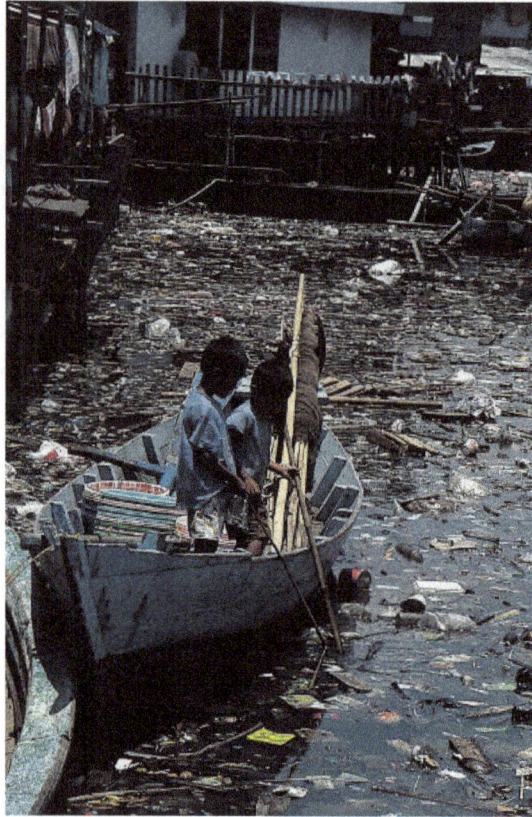

While marine pollution can be obvious, as with the marine debris shown above,
it is often the pollutants that cannot be seen that cause most harm.

Marine pollution occurs when harmful, or potentially harmful, effects result from the entry into the ocean of chemicals, particles, industrial, agricultural and residential waste, noise, or the spread of invasive organisms. Eighty percent of marine pollution comes from land. Air pollution is also a contributing factor by carrying off pesticides or dirt into the ocean. Land and air pollution have proven to be harmful to marine life and its habitats.

The pollution often comes from non point sources such as agricultural runoff, wind-blown debris and dust. Nutrient pollution, a form of water pollution, refers to contamination by excessive inputs of nutrients. It is a primary cause of eutrophication of surface waters, in which excess nutrients, usually nitrogen or phosphorus, stimulate algae growth.

Many potentially toxic chemicals adhere to tiny particles which are then taken up by plankton and benthos animals, most of which are either deposit or filter feeders. In this way, the toxins are concentrated upward within ocean food chains. Many particles combine chemically in a manner highly depletive of oxygen, causing estuaries to become anoxic.

When pesticides are incorporated into the marine ecosystem, they quickly become absorbed into marine food webs. Once in the food webs, these pesticides can cause mutations, as well as diseases, which can be harmful to humans as well as the entire food web.

Toxic metals can also be introduced into marine food webs. These can cause a change to tissue matter, biochemistry, behaviour, reproduction, and suppress growth in marine life. Also, many animal feeds have a high fish meal or fish hydrolysate content. In this way, marine toxins can be transferred to land animals, and appear later in meat and dairy products.

History

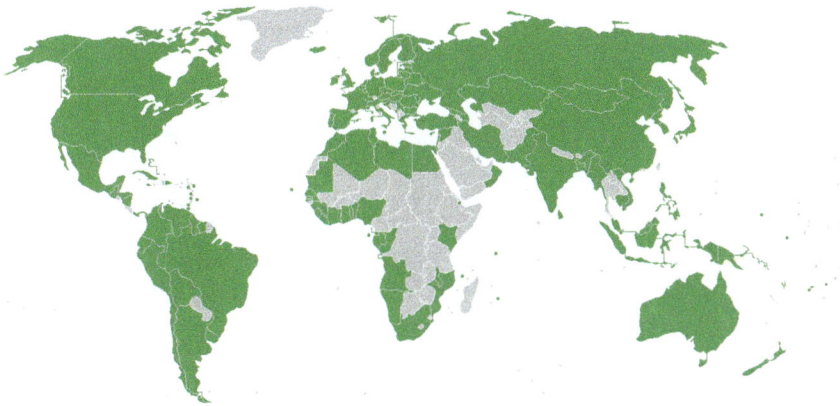

Parties to the MARPOL 73/78 convention on marine pollution

Although marine pollution has a long history, significant international laws to counter it were only enacted in the twentieth century. Marine pollution was a concern during several United Nations Conferences on the Law of the Sea beginning in the 1950s. Most scientists believed that the oceans were so vast that they had unlimited ability to dilute, and thus render pollution harmless.

In the late 1950s and early 1960s, there were several controversies about dumping radioactive waste off the coasts of the United States by companies licensed by the Atomic Energy Commission, into the Irish Sea from the British reprocessing facility at Windscale, and into the Mediterranean Sea by the French Commissariat à l'Energie Atomique. After the Mediterranean Sea controversy, for example, Jacques Cousteau became a worldwide figure in the campaign to stop marine pollution. Marine pollution made further international headlines after the 1967 crash of the oil tanker Torrey Canyon, and after the 1969 Santa Barbara oil spill off the coast of California.

Marine pollution was a major area of discussion during the 1972 United Nations Conference on the Human Environment, held in Stockholm. That year also saw the signing of the Convention on the Prevention of Marine Pollution by Dumping of Wastes and Other Matter, sometimes called the London Convention. The London Convention did not ban marine pollution, but it established black and gray lists for substances to be banned (black) or regulated by national authorities (gray). Cyanide and high-level radioactive waste, for example, were put on the black list. The London Convention applied only to waste dumped from ships, and thus did nothing to regulate waste discharged as liquids from pipelines.

Pathways of Pollution

There are many different ways to categorize, and examine the inputs of pollution into our marine ecosystems. Patin (n.d.) notes that generally there are three main types of inputs of pollution into the ocean: direct discharge of waste into the oceans, runoff into the waters due to rain, and pollutants that are released from the atmosphere.

Septic river.

One common path of entry by contaminants to the sea are rivers. The evaporation of water from oceans exceeds precipitation. The balance is restored by rain over the continents entering rivers and then being returned to the sea. The Hudson in New York State and the Raritan in New Jersey, which empty at the northern and southern ends of Staten Island, are a source of mercury contamination of zooplankton (copepods) in the open ocean. The highest concentration in the filter-feeding copepods is not at the mouths of these rivers but 70 miles south, nearer Atlantic City, because water flows close to the coast. It takes a few days before toxins are taken up by the plankton.

Pollution is often classed as point source or nonpoint source pollution. Point source pollution occurs when there is a single, identifiable, and localized source of the pollution. An example is directly discharging sewage and industrial waste into the ocean. Pollution such as this occurs particularly in developing nations. Nonpoint source pollution occurs when the pollution comes from ill-defined and diffuse sources. These can be difficult to regulate. Agricultural runoff and wind blown debris are prime examples.

Direct Discharge

Acid mine drainage in the Rio Tinto River.

Pollutants enter rivers and the sea directly from urban sewerage and industrial waste discharges, sometimes in the form of hazardous and toxic wastes.

Inland mining for copper, gold. etc., is another source of marine pollution. Most of the pollution is simply soil, which ends up in rivers flowing to the sea. However, some minerals discharged in the course of the mining can cause problems, such as copper, a common industrial pollutant, which can interfere with the life history and development of coral polyps. Mining has a poor environmental track record. For example, according to the United States Environmental Protection Agency, mining has contaminated portions of the headwaters of over 40% of watersheds in the western continental US. Much of this pollution finishes up in the sea.

Land Runoff

Surface runoff from farming, as well as urban runoff and runoff from the construction of roads, buildings, ports, channels, and harbours, can carry soil and particles laden with carbon, nitrogen, phosphorus, and minerals. This nutrient-rich water can cause fleshy algae and phytoplankton to thrive in coastal areas; known as algal blooms, which have the potential to create hypoxic conditions by using all available oxygen.

Polluted runoff from roads and highways can be a significant source of water pollution in coastal areas. About 75% of the toxic chemicals that flow into Puget Sound are carried by stormwater that runs off paved roads and driveways, rooftops, yards and other developed land.

Ship Pollution

A cargo ship pumps ballast water over the side.

Ships can pollute waterways and oceans in many ways. Oil spills can have devastating effects. While being toxic to marine life, polycyclic aromatic hydrocarbons (PAHs), found in crude oil, are very difficult to clean up, and last for years in the sediment and marine environment.

Oil spills are probably the most emotive of marine pollution events. However, while a tanker wreck may result in extensive newspaper headlines, much of the oil in the world's seas comes from other smaller sources, such as tankers discharging ballast water from oil tanks used on return ships, leaking pipelines or engine oil disposed of down sewers.

Discharge of cargo residues from bulk carriers can pollute ports, waterways and oceans. In many instances vessels intentionally discharge illegal wastes despite foreign and domestic regulation prohibiting such actions. It has been estimated that container ships lose over 10,000 containers at sea each year (usually during storms). Ships also create noise pollution that disturbs natural wildlife, and water from ballast tanks can spread harmful algae and other invasive species.

Ballast water taken up at sea and released in port is a major source of unwanted exotic marine life. The invasive freshwater zebra mussels, native to the Black, Caspian and Azov seas, were probably transported to the Great Lakes via ballast water from a transoceanic vessel. Meinesz believes that one of the worst cases of a single invasive species causing harm to an ecosystem can be attributed to a seemingly harmless jellyfish. *Mnemiopsis leidyi*, a species of comb jellyfish that spread so it now inhabits estuaries in many parts of the world. It was first introduced in 1982, and thought to have been transported to the Black Sea in a ship's ballast water. The population of the jellyfish shot up exponentially and, by 1988, it was wreaking havoc upon the local fishing industry. "The anchovy catch fell from 204,000 tons in 1984 to 200 tons in 1993; sprat from 24,600 tons in 1984 to 12,000 tons in 1993; horse mackerel from 4,000 tons in 1984 to zero in 1993." Now that the jellyfish have exhausted the zooplankton, including fish larvae, their numbers have fallen dramatically, yet they continue to maintain a stranglehold on the ecosystem.

Invasive species can take over once occupied areas, facilitate the spread of new diseases, introduce new genetic material, alter underwater seascapes and jeopardize the ability of native species to obtain food. Invasive species are responsible for about $138 billion annually in lost revenue and management costs in the US alone.

Atmospheric Pollution

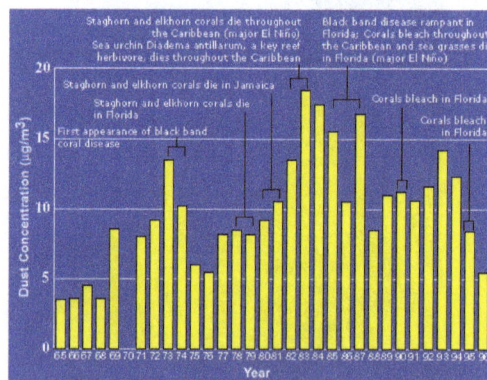

Graph linking atmospheric dust to various coral deaths across the Caribbean Sea and Florida

Another pathway of pollution occurs through the atmosphere. Wind blown dust and debris, including plastic bags, are blown seaward from landfills and other areas. Dust from the Sahara moving around the southern periphery of the subtropical ridge moves into the Caribbean and Florida during the warm season as the ridge builds and moves northward through the subtropical Atlantic. Dust can also be attributed to a global transport from the Gobi and Taklamakan deserts across

Korea, Japan, and the Northern Pacific to the Hawaiian Islands. Since 1970, dust outbreaks have worsened due to periods of drought in Africa. There is a large variability in dust transport to the Caribbean and Florida from year to year; however, the flux is greater during positive phases of the North Atlantic Oscillation. The USGS links dust events to a decline in the health of coral reefs across the Caribbean and Florida, primarily since the 1970s.

Climate change is raising ocean temperatures and raising levels of carbon dioxide in the atmosphere. These rising levels of carbon dioxide are acidifying the oceans. This, in turn, is altering aquatic ecosystems and modifying fish distributions, with impacts on the sustainability of fisheries and the livelihoods of the communities that depend on them. Healthy ocean ecosystems are also important for the mitigation of climate change.

Deep Sea Mining

Deep sea mining is a relatively new mineral retrieval process that takes place on the ocean floor. Ocean mining sites are usually around large areas of polymetallic nodules or active and extinct hydrothermal vents at about 1,400 – 3,700 meters below the ocean's surface. The vents create sulfide deposits, which contain precious metals such as silver, gold, copper, manganese, cobalt, and zinc. The deposits are mined using either hydraulic pumps or bucket systems that take ore to the surface to be processed. As with all mining operations, deep sea mining raises questions about environmental damages to the surrounding areas

Because deep sea mining is a relatively new field, the complete consequences of full scale mining operations are unknown. However, experts are certain that removal of parts of the sea floor will result in disturbances to the benthic layer, increased toxicity of the water column and sediment plumes from tailings. Removing parts of the sea floor disturbs the habitat of benthic organisms, possibly, depending on the type of mining and location, causing permanent disturbances. Aside from direct impact of mining the area, leakage, spills and corrosion would alter the mining area's chemical makeup.

Among the impacts of deep sea mining, sediment plumes could have the greatest impact. Plumes are caused when the tailings from mining (usually fine particles) are dumped back into the ocean, creating a cloud of particles floating in the water. Two types of plumes occur: near bottom plumes and surface plumes. Near bottom plumes occur when the tailings are pumped back down to the mining site. The floating particles increase the turbidity, or cloudiness, of the water, clogging filter-feeding apparatuses used by benthic organisms. Surface plumes cause a more serious problem. Depending on the size of the particles and water currents the plumes could spread over vast areas. The plumes could impact zooplankton and light penetration, in turn affecting the food web of the area.

Types of Pollution

Acidification

The oceans are normally a natural carbon sink, absorbing carbon dioxide from the atmosphere. Because the levels of atmospheric carbon dioxide are increasing, the oceans are becoming more acidic. The potential consequences of ocean acidification are not fully understood, but there are concerns that structures made of calcium carbonate may become vulnerable to dissolution, affecting corals and the ability of shellfish to form shells.

Island with fringing reef in the Maldives. Coral reefs are dying around the world.

Oceans and coastal ecosystems play an important role in the global carbon cycle and have removed about 25% of the carbon dioxide emitted by human activities between 2000 and 2007 and about half the anthropogenic CO_2 released since the start of the industrial revolution. Rising ocean temperatures and ocean acidification means that the capacity of the ocean carbon sink will gradually get weaker, giving rise to global concerns expressed in the Monaco and Manado Declarations.

A report from NOAA scientists published in the journal Science in May 2008 found that large amounts of relatively acidified water are upwelling to within four miles of the Pacific continental shelf area of North America. This area is a critical zone where most local marine life lives or is born. While the paper dealt only with areas from Vancouver to northern California, other continental shelf areas may be experiencing similar effects.

A related issue is the methane clathrate reservoirs found under sediments on the ocean floors. These trap large amounts of the greenhouse gas methane, which ocean warming has the potential to release. In 2004 the global inventory of ocean methane clathrates was estimated to occupy between one and five million cubic kilometres. If all these clathrates were to be spread uniformly across the ocean floor, this would translate to a thickness between three and fourteen metres. This estimate corresponds to 500–2500 gigatonnes carbon (Gt C), and can be compared with the 5000 Gt C estimated for all other fossil fuel reserves.

Eutrophication

Eutrophication is an increase in chemical nutrients, typically compounds containing nitrogen or phosphorus, in an ecosystem. It can result in an increase in the ecosystem's primary productivity (excessive plant growth and decay), and further effects including lack of oxygen and severe reductions in water quality, fish, and other animal populations.

The biggest culprit are rivers that empty into the ocean, and with it the many chemicals used as fertilizers in agriculture as well as waste from livestock and humans. An excess of oxygen depleting chemicals in the water can lead to hypoxia and the creation of a dead zone.

Polluted lagoon.

Estuaries tend to be naturally eutrophic because land-derived nutrients are concentrated where runoff enters the marine environment in a confined channel. The World Resources Institute has identified 375 hypoxic coastal zones around the world, concentrated in coastal areas in Western Europe, the Eastern and Southern coasts of the US, and East Asia, particularly in Japan. In the ocean, there are frequent red tide algae blooms that kill fish and marine mammals and cause respiratory problems in humans and some domestic animals when the blooms reach close to shore.

Effect of eutrophication on marine benthic life

In addition to land runoff, atmospheric anthropogenic fixed nitrogen can enter the open ocean. A study in 2008 found that this could account for around one third of the ocean's external (non-recycled) nitrogen supply and up to three per cent of the annual new marine biological production. It has been suggested that accumulating reactive nitrogen in the environment may have consequences as serious as putting carbon dioxide in the atmosphere.

One proposed solution to eutrophication in estuaries is to restore shellfish populations, such as oysters. Oyster reefs remove nitrogen from the water column and filter out suspended solids, subsequently reducing the likelihood or extent of harmful algal blooms or anoxic conditions. Filter feeding activity is considered beneficial to water quality by controlling phytoplankton density and sequestering nutrients, which can be removed from the system through shellfish harvest, buried in the sediments, or lost through denitrification. Foundational work toward the idea of improving marine water quality through shellfish cultivation to was conducted by Odd Lindahl et al., using mussels in Sweden.

Plastic Debris

A mute swan builds a nest using plastic garbage.

Marine debris is mainly discarded human rubbish which floats on, or is suspended in the ocean. Eighty percent of marine debris is plastic – a component that has been rapidly accumulating since the end of World War II. The mass of plastic in the oceans may be as high as 100,000,000 tonnes (98,000,000 long tons; 110,000,000 short tons).

Discarded plastic bags, six pack rings and other forms of plastic waste which finish up in the ocean present dangers to wildlife and fisheries. Aquatic life can be threatened through entanglement, suffocation, and ingestion. Fishing nets, usually made of plastic, can be left or lost in the ocean by fishermen. Known as ghost nets, these entangle fish, dolphins, sea turtles, sharks, dugongs, crocodiles, seabirds, crabs, and other creatures, restricting movement, causing starvation, laceration and infection, and, in those that need to return to the surface to breathe, suffocation.

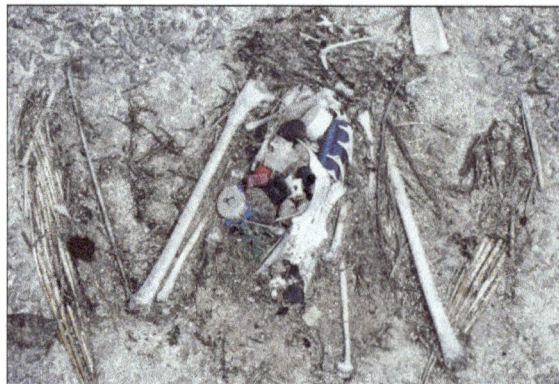

Remains of an albatross containing ingested flotsam

Many animals that live on or in the sea consume flotsam by mistake, as it often looks similar to their natural prey. Plastic debris, when bulky or tangled, is difficult to pass, and may become permanently lodged in the digestive tracts of these animals. Especially when evolutionary adaptions make it impossible for the likes of turtles to reject plastic bags, which resemble jellyfish when immersed in water, as they have a system in their throat to stop slippery foods from otherwise escaping. Thereby blocking the passage of food and causing death through starvation or infection.

Plastics accumulate because they don't biodegrade in the way many other substances do. They will photodegrade on exposure to the sun, but they do so properly only under dry conditions, and water inhibits this process. In marine environments, photodegraded plastic disintegrates into ever smaller pieces while remaining polymers, even down to the molecular level. When floating plastic particles photodegrade down to zooplankton sizes, jellyfish attempt to consume them, and in this way the plastic enters the ocean food chain. Many of these long-lasting pieces end up in the stomachs of marine birds and animals, including sea turtles, and black-footed albatross.

Marine debris on Kamilo Beach, Hawaii, washed up from the Great Pacific Garbage Patch

Plastic debris tends to accumulate at the centre of ocean gyres. In particular, the Great Pacific Garbage Patch has a very high level of plastic particulate suspended in the upper water column. In samples taken in 1999, the mass of plastic exceeded that of zooplankton (the dominant animal life in the area) by a factor of six. Midway Atoll, in common with all the Hawaiian Islands, receives substantial amounts of debris from the garbage patch. Ninety percent plastic, this debris accumulates on the beaches of Midway where it becomes a hazard to the bird population of the island. Midway Atoll is home to two-thirds (1.5 million) of the global population of Laysan albatross. Nearly all of these albatross have plastic in their digestive system and one-third of their chicks die.

Toxic additives used in the manufacture of plastic materials can leach out into their surroundings when exposed to water. Waterborne hydrophobic pollutants collect and magnify on the surface of plastic debris, thus making plastic far more deadly in the ocean than it would be on land. Hydrophobic contaminants are also known to bioaccumulate in fatty tissues, biomagnifying up the food chain and putting pressure on apex predators. Some plastic additives are known to disrupt the endocrine system when consumed, others can suppress the immune system or decrease reproductive rates. Floating debris can also absorb persistent organic pollutants from seawater, including PCBs, DDT and PAHs. Aside from toxic effects, when ingested some of these are mistaken by the animal brain for estradiol, causing hormone disruption in the affected wildlife.

Toxins

Apart from plastics, there are particular problems with other toxins that do not disintegrate rapidly in the marine environment. Examples of persistent toxins are PCBs, DDT, TBT, pesticides, furans, dioxins, phenols and radioactive waste. Heavy metals are metallic chemical elements that have a relatively high density and are toxic or poisonous at low concentrations. Examples are mercury, lead, nickel, arsenic and cadmium. Such toxins can accumulate in the tissues of many species of aquatic life in a process called bioaccumulation. They are also known to accumulate in benthic environments, such as estuaries and bay muds: a geological record of human activities of the last century.

Specific Examples

- Chinese and Russian industrial pollution such as phenols and heavy metals in the Amur River have devastated fish stocks and damaged its estuary soil.

- Wabamun Lake in Alberta, Canada, once the best whitefish lake in the area, now has unacceptable levels of heavy metals in its sediment and fish.

- Acute and chronic pollution events have been shown to impact southern California kelp forests, though the intensity of the impact seems to depend on both the nature of the contaminants and duration of exposure.

- Due to their high position in the food chain and the subsequent accumulation of heavy metals from their diet, mercury levels can be high in larger species such as bluefin and albacore. As a result, in March 2004 the United States FDA issued guidelines recommending that pregnant women, nursing mothers and children limit their intake of tuna and other types of predatory fish.

- Some shellfish and crabs can survive polluted environments, accumulating heavy metals or toxins in their tissues. For example, mitten crabs have a remarkable ability to survive in highly modified aquatic habitats, including polluted waters. The farming and harvesting of such species needs careful management if they are to be used as a food.

- Surface runoff of pesticides can alter the gender of fish species genetically, transforming male into female fish.

- Heavy metals enter the environment through oil spills – such as the Prestige oil spill on the Galician coast – or from other natural or anthropogenic sources.

- In 2005, the 'Ndrangheta, an Italian mafia syndicate, was accused of sinking at least 30 ships loaded with toxic waste, much of it radioactive. This has led to widespread investigations into radioactive-waste disposal rackets.

- Since the end of World War II, various nations, including the Soviet Union, the United Kingdom, the United States, and Germany, have disposed of chemical weapons in the Baltic Sea, raising concerns of environmental contamination.

Underwater Noise

Marine life can be susceptible to noise or the sound pollution from sources such as passing ships, oil exploration seismic surveys, and naval low-frequency active sonar. Sound travels more rapidly and over larger distances in the sea than in the atmosphere. Marine animals, such as cetaceans, often have weak eyesight, and live in a world largely defined by acoustic information. This applies also to many deeper sea fish, who live in a world of darkness. Between 1950 and 1975, ambient noise at one location in the Pacific Ocean increased by about ten decibels (that is a tenfold increase in intensity).

Noise also makes species communicate louder, which is called the Lombard vocal response. Whale songs are longer when submarine-detectors are on. If creatures don't "speak" loud enough, their voice can be masked by anthropogenic sounds. These unheard voices might be warnings, finding of prey, or preparations of net-bubbling. When one species begins speaking louder, it will mask other species voices, causing the whole ecosystem to eventually speak louder.

According to the oceanographer Sylvia Earle, "Undersea noise pollution is like the death of a thousand cuts. Each sound in itself may not be a matter of critical concern, but taken all together, the noise from shipping, seismic surveys, and military activity is creating a totally different environment than existed even 50 years ago. That high level of noise is bound to have a hard, sweeping impact on life in the sea."

Adaptation and Mitigation

Aerosol can polluting a beach.

Much anthropogenic pollution ends up in the ocean. The 2011 edition of the United Nations Environment Programme Year Book identifies as the main emerging environmental issues the loss to the oceans of massive amounts of phosphorus, "a valuable fertilizer needed to feed a growing global population", and the impact billions of pieces of plastic waste are having globally on the health of marine environments. Bjorn Jenssen (2003) notes in his article, "Anthropogenic pollution may reduce biodiversity and productivity of marine ecosystems, resulting in reduction and depletion of human marine food resources". There are two ways the overall level of this pollution can be mitigated: either the human population is reduced, or a way is found to reduce the ecological footprint left behind by the average human. If the second way is not adopted, then the first way may be imposed as world ecosystems falter.

The second way is for humans, individually, to pollute less. That requires social and political will, together with a shift in awareness so more people respect the environment and are less disposed to abuse it. At an operational level, regulations, and international government participation is needed. It is often very difficult to regulate marine pollution because pollution spreads over international barriers, thus making regulations hard to create as well as enforce.

Without appropriate awareness of marine pollution, the necessary global will to effectively address the issues may prove inadequate. Balanced information on the sources and harmful effects of marine pollution need to become part of general public awareness, and ongoing research is required to fully establish, and keep current, the scope of the issues. As expressed in Daoji and Dag's research, one of the reasons why environmental concern is lacking among the Chinese is because the public awareness is low and therefore should be targeted. Likewise, regulation, based upon such in-depth research should be employed. In California, such regulations have already been put in place to protect Californian coastal waters from agricultural runoff. This includes the California Water Code, as well as several voluntary programs. Similarly, in India, several tactics have been employed that help reduce marine pollution, however, they do not significantly target the problem. In Chennai, sewage has been dumped further into open waters. Due to the mass of waste being deposited, open-ocean is best for diluting, and dispersing pollutants, thus making them less harmful to marine ecosystems.

Ocean Acidification

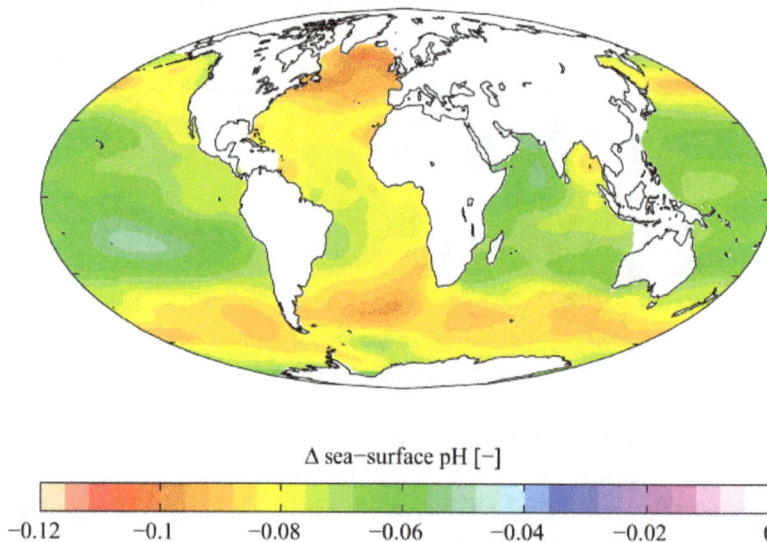

Δ sea–surface pH [–]

```
−0.12      −0.1      −0.08      −0.06      −0.04      −0.02       0
```

Estimated change in sea water pH caused by human created CO2 between the 1700s and the 1990s, from the Global Ocean Data Analysis Project (GLODAP) and the World Ocean Atlas

Ocean acidification is the ongoing decrease in the pH of the Earth's oceans, caused by the uptake of carbon dioxide (CO_2) from the atmosphere. Seawater is slightly basic (meaning pH > 7), and the process in question is a shift towards pH-neutral conditions rather than a transition to acidic conditions (pH < 7). Ocean alkalinity is not changed by the process, or may increase over long time periods due to carbonate dissolution. An estimated 30–40% of the carbon dioxide from human activity released into the atmosphere dissolves into oceans, rivers and lakes. To achieve chemical equilibrium, some of it reacts with the water to form carbonic acid. Some of these extra carbonic acid molecules react with a water molecule to give a bicarbonate ion and a hydronium ion, thus increasing ocean acidity (H^+ ion concentration). Between 1751 and 1994 surface ocean pH is estimated to have decreased from approximately 8.25 to 8.14, representing an increase of almost 30% in H^+ ion concentration in the world's oceans. Earth System Models project that within the last decade ocean acidity exceeded historical analogs and in combination with other ocean biogeochemical changes could undermine the functioning of marine ecosystems and disrupt the provision of many goods and services associated with the ocean.

Increasing acidity is thought to have a range of potentially harmful consequences for marine organisms, such as depressing metabolic rates and immune responses in some organisms, and causing coral bleaching. By increasing the presence of free hydrogen ions, each molecule of carbonic acid that forms in the oceans ultimately results in the conversion of *two* carbonate ions into bicarbonate ions. This net decrease in the amount of carbonate ions available makes it more difficult for marine calcifying organisms, such as coral and some plankton, to form biogenic calcium carbonate, and such structures become vulnerable to dissolution. Ongoing acidification of the oceans threatens food chains connected with the oceans. As members of the InterAcademy Panel, 105 science academies have issued a statement on ocean acidification recommending that by 2050, global CO_2 emissions be reduced by at least 50% compared to the 1990 level.

While ongoing ocean acidification is anthropogenic in origin, it has occurred previously in Earth's history. The most notable example is the Paleocene-Eocene Thermal Maximum (PETM), which occurred approximately 56 million years ago. For reasons that are currently uncertain, massive amounts of carbon entered the ocean and atmosphere, and led to the dissolution of carbonate sediments in all ocean basins.

Ocean acidification has been called the "evil twin of global warming" and "the other CO_2 problem".

Carbon Cycle

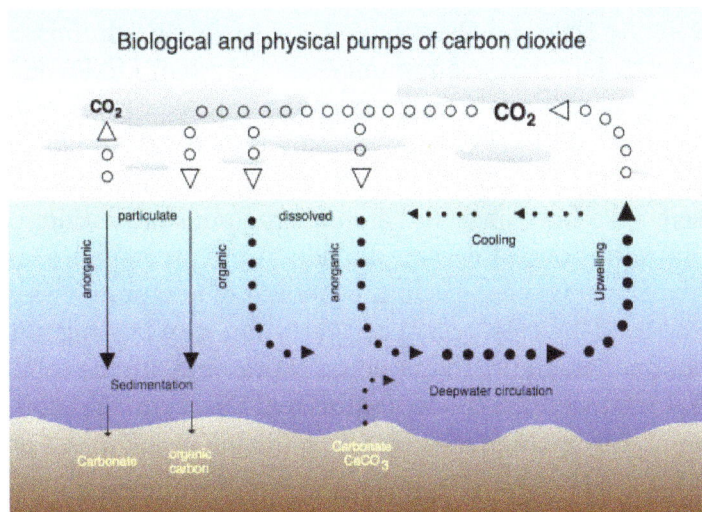

The CO2 cycle between the atmosphere and the ocean

The carbon cycle describes the fluxes of carbon dioxide (CO2) between the oceans, terrestrial biosphere, lithosphere, and the atmosphere. Human activities such as the combustion of fossil fuels and land use changes have led to a new flux of CO2 into the atmosphere. About 45% has remained in the atmosphere; most of the rest has been taken up by the oceans, with some taken up by terrestrial plants.

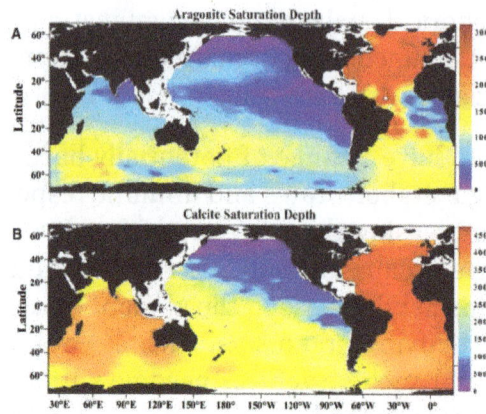

Distribution of (A) aragonite and (B) calcite saturation depth in the global oceans

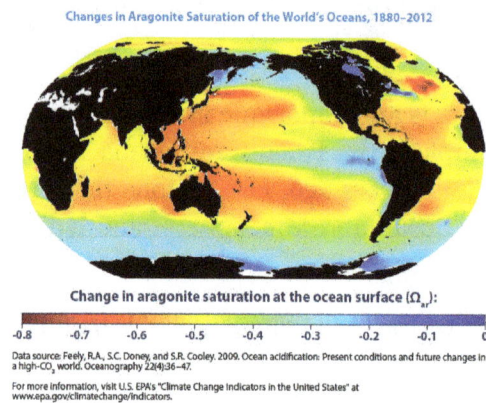

Changes in Aragonite Saturation of the World's Oceans, 1880–2012

Change in aragonite saturation at the ocean surface (Ω_a):

| -0.8 | -0.7 | -0.6 | -0.5 | -0.4 | -0.3 | -0.2 | -0.1 | 0 |

Data source: Feely, R.A., S.C. Doney, and S.R. Cooley. 2009. Ocean acidification: Present conditions and future changes in a high-CO_2 world. Oceanography 22(4):36–47.

For more information, visit U.S. EPA's "Climate Change Indicators in the United States" at www.epa.gov/climatechange/indicators.

The map was created by the National Oceanic and Atmospheric Administration and the Woods Hole Oceanographic Institution using Community Earth System Model data. This map was created by comparing average conditions during the 1880s with average conditions during the most recent 10 years (2003–2012). Aragonite saturation has only been measured at selected locations during the last few decades, but it can be calculated reliably for different times and locations based on the relationships scientists have observed among aragonite saturation, pH, dissolved carbon, water temperature, concentrations of carbon dioxide in the atmosphere, and other factors that can be measured. This map shows changes in the amount of aragonite dissolved in ocean surface waters between the 1880s and the most recent decade (2003–2012). Aragonite saturation is a ratio that compares the amount of aragonite that is actually present with the total amount of aragonite that the water could hold if it were completely saturated. The more negative the change in aragonite saturation, the larger the decrease in aragonite available in the water, and the harder it is for marine creatures to produce their skeletons and shells. The global map shows changes over time in the amount of aragonite dissolved in ocean water, which is called aragonite saturation.

The carbon cycle involves both organic compounds such as cellulose and inorganic carbon compounds such as carbon dioxide and the carbonates. The inorganic compounds are particularly relevant when discussing ocean acidification for it includes many forms of dissolved CO_2 present in the Earth's oceans.

When CO_2 dissolves, it reacts with water to form a balance of ionic and non-ionic chemical species: dissolved free carbon dioxide (CO2(aq)), carbonic acid (H2CO3), bicarbonate (HCO−3) and carbonate (CO2−3). The ratio of these species depends on factors such as seawater temperature and alkalinity (as shown in a Bjerrum plot). These different forms of dissolved inorganic carbon are transferred from an ocean's surface to its interior by the ocean's solubility pump.

The resistance of an area of ocean to absorbing atmospheric CO2 is known as the Revelle factor.

Acidification

Dissolving CO2 in seawater increases the hydrogen ion (H+) concentration in the ocean, and thus decreases ocean pH, as follows:

$$CO_{2\,(aq)} + H_2O\ H_2CO_3\ HCO_3^- + H^+\ CO_3^{2-} + 2\ H^+.$$

Caldeira and Wickett (2003) placed the rate and magnitude of modern ocean acidification changes in the context of probable historical changes during the last 300 million years.

Since the industrial revolution began, it is estimated that surface ocean pH has dropped by slightly more than 0.1 units on the logarithmic scale of pH, representing about a 29% increase in H+ . It is expected to drop by a further 0.3 to 0.5 pH units (an additional doubling to tripling of today's post-industrial acid concentrations) by 2100 as the oceans absorb more anthropogenic CO 2, the impacts being most severe for coral reefs and the Southern Ocean. These changes are predicted to continue rapidly as the oceans take up more anthropogenic CO 2 from the atmosphere. The degree of change to ocean chemistry, including ocean pH, will depend on the mitigation and emissions pathways society takes.

Although the largest changes are expected in the future, a report from NOAA scientists found large quantities of water undersaturated in aragonite are already upwelling close to the Pacific continental shelf area of North America. Continental shelves play an important role in marine ecosystems since most marine organisms live or are spawned there, and though the study only dealt with the area from Vancouver to Northern California, the authors suggest that other shelf areas may be experiencing similar effects.

Average surface ocean pH				
Time	pH	pH change relative to pre-industrial	Source	H+ concentration change relative to pre-industrial
Pre-industrial (18th century)	8.179		analysed field	
Recent past (1990s)	8.104	−0.075	field	+ 18.9%
Present levels	~8.069	−0.11	field	+ 28.8%
2050 (2×CO 2 = 560 ppm)	7.949	−0.230	model	+ 69.8%
2100 (IS92a)	7.824	−0.355	model	+ 126.5%

Rate

One of the first detailed datasets to examine how pH varied over a period of time at a temperate coastal location found that acidification was occurring much faster than previously predicted, with consequences for near-shore benthic ecosystems. Thomas Lovejoy, former chief biodiversity advisor to the World Bank, has suggested that "the acidity of the oceans will more than double in the next 40 years. This rate is 100 times faster than any changes in ocean acidity in the last 20 million years, making it unlikely that marine life can somehow adapt to the changes." It is predicted that, by the year 2100, the level of acidity in the ocean will reach the levels experienced by the earth 20 million years ago.

Current rates of ocean acidification have been compared with the greenhouse event at the Paleocene–Eocene boundary (about 55 million years ago) when surface ocean temperatures rose by 5–6 degrees Celsius. No catastrophe was seen in surface ecosystems, yet bottom-dwelling organisms in the deep ocean experienced a major extinction. The current acidification is on a path to reach levels higher than any seen in the last 65 million years, and the rate of increase is about ten times the rate that preceded the Paleocene–Eocene mass extinction. The current and projected acidification has been described as an almost unprecedented geological event. A National Research Council study released in April 2010 likewise concluded that "the level of acid in the oceans is increasing at an unprecedented rate." A 2012 paper in the journal *Science* examined the geological record in an attempt to find a historical analog for current global conditions as well as those of the future. The researchers determined that the current rate of ocean acidification is faster than at any time in the past 300 million years.

A review by climate scientists at the RealClimate blog, of a 2005 report by the Royal Society of the UK similarly highlighted the centrality of the *rates* of change in the present anthropogenic acidification process, writing:

"The natural pH of the ocean is determined by a need to balance the deposition and burial of $CaCO_3$ on the sea floor against the influx of $Ca2+$ and $CO2-3$ into the ocean from dissolving rocks on land, called weathering. These processes stabilize the pH of the ocean, by a mechanism called $CaCO_3$ compensation...The point of bringing it up again is to note that if the CO_2 concentration of the atmosphere changes more slowly than this, as it always has throughout the Vostok record, the pH of the ocean will be relatively unaffected because $CaCO3$ compensation can keep up. The [present] fossil fuel acidification is much faster than natural changes, and so the acid spike will be more intense than the earth has seen in at least 800,000 years."

In the 15-year period 1995–2010 alone, acidity has increased 6 percent in the upper 100 meters of the Pacific Ocean from Hawaii to Alaska. According to a statement in July 2012 by Jane Lubchenco, head of the U.S. National Oceanic and Atmospheric Administration "surface waters are changing much more rapidly than initial calculations have suggested. It's yet another reason to be very seriously concerned about the amount of carbon dioxide that is in the atmosphere now and the additional amount we continue to put out."

A 2013 study claimed acidity was increasing at a rate 10 times faster than in any of the evolutionary crises in Earth's history. In a synthesis report published in *Science* in 2015, 22 leading marine scientists stated that CO_2 from burning fossil fuels is changing the oceans' chemistry more rapidly than at any time since the Great Dying, Earth's most severe known extinction event, emphasizing that the 2 °C maximum temperature increase agreed upon by governments reflects too small a cut in emissions to prevent "dramatic impacts" on the world's oceans, with lead author Jean-Pierre

Gattuso remarking that "The ocean has been minimally considered at previous climate negotiations. Our study provides compelling arguments for a radical change at the UN conference (in Paris) on climate change".

Calcification

Overview

Changes in ocean chemistry can have extensive direct and indirect effects on organisms and their habitats. One of the most important repercussions of increasing ocean acidity relates to the production of shells and plates out of calcium carbonate ($CaCO_3$). This process is called calcification and is important to the biology and survival of a wide range of marine organisms. Calcification involves the precipitation of dissolved ions into solid $CaCO_3$ structures, such as coccoliths. After they are formed, such structures are vulnerable to dissolution unless the surrounding seawater contains saturating concentrations of carbonate ions (CO_3^{2-}).

Mechanism

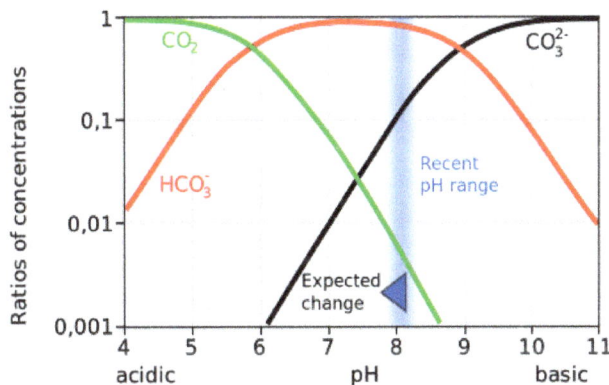

Bjerrum plot: Change in carbonate system of seawater from ocean acidification.

Of the extra carbon dioxide added into the oceans, some remains as dissolved carbon dioxide, while the rest contributes towards making additional bicarbonate (and additional carbonic acid). This also increases the concentration of hydrogen ions, and the percentage increase in hydrogen is larger than the percentage increase in bicarbonate, creating an imbalance in the reaction $HCO_3^- \rightleftharpoons CO_3^{2-} + H^+$. To maintain chemical equilibrium, some of the carbonate ions already in the ocean combine with some of the hydrogen ions to make further bicarbonate. Thus the ocean's concentration of carbonate ions is reduced, creating an imbalance in the reaction $Ca^{2+} + CO_3^{2-} \rightleftharpoons CaCO_3$, and making the dissolution of formed $CaCO_3$ structures more likely.

These increases in concentrations of dissolved carbon dioxide and bicarbonate, and reduction in carbonate, are shown in a Bjerrum plot.

Saturation State

The saturation state (known as Ω) of seawater for a mineral is a measure of the thermodynamic potential for the mineral to form or to dissolve, and is described by the following equation:

Here Ω is the product of the concentrations (or activities) of the reacting ions that form the mineral (Ca2+ and CO2−3), divided by the product of the concentrations of those ions when the mineral is at equilibrium (Ksp), that is, when the mineral is neither forming nor dissolving. In seawater, a natural horizontal boundary is formed as a result of temperature, pressure, and depth, and is known as the saturation horizon, or lysocline. Above this saturation horizon, Ω has a value greater than 1, and CaCO3 does not readily dissolve. Most calcifying organisms live in such waters. Below this depth, Ω has a value less than 1, and CaCO3 will dissolve. However, if its production rate is high enough to offset dissolution, CaCO3 can still occur where Ω is less than 1. The carbonate compensation depth occurs at the depth in the ocean where production is exceeded by dissolution.

The decrease in the concentration of CO_3^{2-} decreases Ω, and hence makes CaCO3 dissolution more likely.

Calcium carbonate occurs in two common polymorphs (crystalline forms): aragonite and calcite. Aragonite is much more soluble than calcite, so the aragonite saturation horizon is always nearer to the surface than the calcite saturation horizon. This also means that those organisms that produce aragonite may be more vulnerable to changes in ocean acidity than those that produce calcite. Increasing CO2 levels and the resulting lower pH of seawater decreases the saturation state of CaCO3 and raises the saturation horizons of both forms closer to the surface. This decrease in saturation state is believed to be one of the main factors leading to decreased calcification in marine organisms, as the inorganic precipitation of CaCO3 is directly proportional to its saturation state.

Possible Impacts

Increasing acidity has possibly harmful consequences, such as depressing metabolic rates in jumbo squid, depressing the immune responses of blue mussels, and coral bleaching. However it may benefit some species, for example increasing the growth rate of the sea star, *Pisaster ochraceus*, while shelled plankton species may flourish in altered oceans.

The report "Ocean Acidification Summary for Policymakers 2013" describes research findings and possible impacts.

Impacts on Oceanic Calcifying Organisms

Although the natural absorption of CO2 by the world's oceans helps mitigate the climatic effects of anthropogenic emissions of CO2, it is believed that the resulting decrease in pH will have negative consequences, primarily for oceanic calcifying organisms. These span the food chain from autotrophs to heterotrophs and include organisms such as coccolithophores, corals, foraminifera, echinoderms, crustaceans and molluscs. As described above, under normal conditions, calcite and aragonite are stable in surface waters since the carbonate ion is at supersaturating concentrations. However, as ocean pH falls, the concentration of carbonate ions required for saturation to occur increases, and when carbonate becomes undersaturated, structures made of calcium carbonate are vulnerable to dissolution. Therefore, even if there is no change in the rate of calcification, the rate of dissolution of calcareous material increases.

Corals, coccolithophore algae, coralline algae, foraminifera, shellfish and pteropods experience reduced calcification or enhanced dissolution when exposed to elevated CO2.

The Royal Society published a comprehensive overview of ocean acidification, and its potential consequences, in June 2005. However, some studies have found different response to ocean acidification, with coccolithophore calcification and photosynthesis both increasing under elevated atmospheric pCO_2, an equal decline in primary production and calcification in response to elevated CO_2 or the direction of the response varying between species. A study in 2008 examining a sediment core from the North Atlantic found that while the species composition of coccolithophorids has remained unchanged for the industrial period 1780 to 2004, the calcification of coccoliths has increased by up to 40% during the same time. A 2010 study from Stony Brook University suggested that while some areas are overharvested and other fishing grounds are being restored, because of ocean acidification it may be impossible to bring back many previous shellfish populations. While the full ecological consequences of these changes in calcification are still uncertain, it appears likely that many calcifying species will be adversely affected.

When exposed in experiments to pH reduced by 0.2 to 0.4, larvae of a temperate brittlestar, a relative of the common sea star, fewer than 0.1 percent survived more than eight days. There is also a suggestion that a decline in the coccolithophores may have secondary effects on climate, contributing to global warming by decreasing the Earth's albedo via their effects on oceanic cloud cover. All marine ecosystems on Earth will be exposed to changes in acidification and several other ocean biogeochemical changes.

The fluid in the internal compartments where corals grow their exoskeleton is also extremely important for calcification growth. When the saturation rate of aragonite in the external seawater is at ambient levels, the corals will grow their aragonite crystals rapidly in their internal compartments, hence their exoskeleton grows rapidly. If the level of aragonite in the external seawater is lower than the ambient level, the corals have to work harder to maintain the right balance in the internal compartment. When that happens, the process of growing the crystals slows down, and this slows down the rate of how much their exoskeleton is growing. Depending on how much aragonite is in the surrounding water, the corals may even stop growing because the levels of aragonite are too low to pump in to the internal compartment. They could even dissolve faster than they can make the crystals to their skeleton, depending on the aragonite levels in the surrounding water.

Ocean acidification may force some organisms to reallocate resources away from productive endpoints such as growth in order to maintain calcification.

In some places carbon dioxide bubbles out from the sea floor, locally changing the pH and other aspects of the chemistry of the seawater. Studies of these carbon dioxide seeps have documented a variety of responses by different organisms. Coral reef communities located near carbon dioxide seeps are of particular interest because of the sensitivity of some corals species to acidification. In Papua New Guinea, declining pH caused by carbon dioxide seeps is associated with declines in coral species diversity. However, in Palau carbon dioxide seeps are not associated with reduced species diversity of corals, although bioerosion of coral skeletons is much higher at low pH sites.

Other Biological Impacts

Aside from the slowing and/or reversing of calcification, organisms may suffer other adverse effects, either indirectly through negative impacts on food resources, or directly as reproductive or physiological effects. For example, the elevated oceanic levels of CO_2 may produce CO2-induced acidification of body fluids, known as hypercapnia. Also, increasing ocean acidity is believed to have a range of direct consequences. For example, increasing acidity has been observed to: reduce metabolic rates in jumbo squid; depress the immune responses of blue mussels; and make it harder for juvenile clownfish to tell apart the smells of non-predators and predators, or hear the sounds of their predators. This is possibly because ocean acidification may alter the acoustic properties of seawater, allowing sound to propagate further, and increasing ocean noise. Calcium carbonate ions are very important when it comes to building organisms as they use calcium to build skeletons and shells. OA affects ocean species to a varying degree. Many plants do well with high CO_2 levels, but organisms that calcify like clams, mussels, corals and sea urchins might not do so well because everything in the ocean is connected including the food web. Since OA is such a large event because the ocean takes up 71% of the earth's surface OA is being referred to as climate change because it will affect the whole planet if the ocean becomes acidic. This impacts all animals that use sound for echolocation or communication. Atlantic longfin squid eggs took longer to hatch in acidified water, and the squid's statolith was smaller and malformed in animals placed in sea water with a lower pH. The lower PH was simulated with 20-30 times the normal amount of CO_2. However, as with calcification, as yet there is not a full understanding of these processes in marine organisms or ecosystems.

Another possible effect would be an increase in red tide events, which could contribute to the accumulation of toxins (domoic acid, brevetoxin, saxitoxin) in small organisms such as anchovies and shellfish, in turn increasing occurrences of amnesic shellfish poisoning, neurotoxic shellfish poisoning and paralytic shellfish poisoning.

Nonbiological Impacts

Leaving aside direct biological effects, it is expected that ocean acidification in the future will lead to a significant decrease in the burial of carbonate sediments for several centuries, and even the dissolution of existing carbonate sediments. This will cause an elevation of ocean alkalinity, leading to the enhancement of the ocean as a reservoir for CO_2 with implications for climate change as more CO_2 leaves the atmosphere for the ocean.

Impact on Human Industry

The threat of acidification includes a decline in commercial fisheries and in the Arctic tourism industry and economy. Commercial fisheries are threatened because acidification harms calcifying organisms which form the base of the Arctic food webs.

Pteropods and brittle stars both form the base of the Arctic food webs and are both seriously damaged from acidification. Pteropods shells dissolve with increasing acidification and the brittle stars lose muscle mass when re-growing appendages. For pteropods to create shells they require aragonite which is produced through carbonate ions and dissolved calcium. Pteropods are severely affected because increasing acidification levels have steadily decreased the amount of water supersaturated with carbonate which is needed for aragonite creation. Arctic waters are changing so rapidly that they will become undersaturated with aragonite as early as 2016. Additionally the brittle star's eggs die within a few days when exposed to expected conditions resulting from Arctic acidification. Acidification threatens to destroy Arctic food webs from the base up. Arctic food webs are considered simple, meaning there are few steps in the food chain from small organisms to larger predators. For example, pteropods are "a key prey item of a number of higher predators - larger plankton, fish, seabirds, whales" Both pteropods and sea stars serve as a substantial food source and their removal from the simple food web would pose a serious threat to the whole ecosystem. The effects on the calcifying organisms at the base of the food webs could potentially destroy fisheries. The value of fish caught from US commercial fisheries in 2007 was valued at $3.8 billion and of that 73% was derived from calcifiers and their direct predators. Other organisms are directly harmed as a result of acidification. For example, decrease in the growth of marine calcifiers such as the American Lobster, Ocean Quahog, and scallops means there is less shellfish meat available for sale and consumption. Red king crab fisheries are also at a serious threat because crabs are calcifiers and rely on carbonate ions for shell development. Baby red king crab when exposed to increased acidification levels experienced 100% mortality after 95 days. In 2006 Red King Cab accounted for 23% of the total guideline harvest levels and a serious decline in red crab population would threaten the crab harvesting industry. Several ocean goods and services are likely to be undermined by future ocean acidification potentially affecting the livelihoods of some 400 to 800 million people depending upon the emission scenario.

Impact on Indigenous Peoples

Acidification could damage the Arctic tourism economy and affect the way of life of indigenous peoples. A major pillar of Arctic tourism is the sport fishing and hunting industry. The sport fishing industry is threatened by collapsing food webs which provide food for the prized fish. A decline in tourism lowers revenue input in the area, and threatens the economies that are increasingly dependent on tourism. Acidification is not merely a threat but has significantly declined whole fish populations. For example, In Scandinavia studies conducted on acidic water revealed that 15% of species populations had disappeared and that many more populations were limited in numbers or declining. The rapid decrease or disappearance of marine life could also affect the diet of Indigenous peoples.

Possible Responses

Reducing CO_2 Emissions

Members of the InterAcademy Panel recommended that by 2050, global anthropogenic CO_2 emissions be reduced less than 50% of the 1990 level. The 2009 statement also called on world leaders to:

- Acknowledge that ocean acidification is a direct and real consequence of increasing atmospheric CO_2 concentrations, is already having an effect at current concentrations, and is likely to cause grave harm to important marine ecosystems as CO_2 concentrations reach 450 [parts-per-million (ppm)] and above;

- [...] Recognise that reducing the build up of CO_2 in the atmosphere is the only practicable solution to mitigating ocean acidification;

- [...] Reinvigorate action to reduce stressors, such as overfishing and pollution, on marine ecosystems to increase resilience to ocean acidification.

Stabilizing atmospheric CO_2 concentrations at 450 ppm would require near-term emissions reductions, with steeper reductions over time.

The German Advisory Council on Global Change stated:

In order to prevent disruption of the calcification of marine organisms and the resultant risk of fundamentally altering marine food webs, the following guard rail should be obeyed: the pH of near surface waters should not drop more than 0.2 units below the pre-industrial average value in any larger ocean region (nor in the global mean).

One policy target related to ocean acidity is the magnitude of future global warming. Parties to the United Nations Framework Convention on Climate Change (UNFCCC) adopted a target of limiting warming to below 2 °C, relative to the pre-industrial level. Meeting this target would require substantial reductions in anthropogenic CO_2 emissions.

Limiting global warming to below 2 °C would imply a reduction in surface ocean pH of 0.16 from pre-industrial levels. This would represent a substantial decline in surface ocean pH.

Climate Engineering

Climate engineering (mitigating temperature or pH effects of emissions) has been proposed as a possible response to ocean acidification. The IAP (2009) statement cautioned against climate engineering as a policy response:

Mitigation approaches such as adding chemicals to counter the effects of acidification are likely to be expensive, only partly effective and only at a very local scale, and may pose additional unanticipated risks to the marine environment. There has been very little research on the feasibility and impacts of these approaches. Substantial research is needed before these techniques could be applied.

Reports by the WGBU (2006), the UK's Royal Society (2009), and the US National Research Council (2011) warned of the potential risks and difficulties associated with climate engineering.

Iron Fertilization

Iron fertilization of the ocean could stimulate photosynthesis in phytoplankton. The phytoplankton would convert the ocean's dissolved carbon dioxide into carbohydrate and oxygen gas, some of which would sink into the deeper ocean before oxidizing. More than a dozen open-sea experiments confirmed that adding iron to the ocean increases photosynthesis in phytoplankton by up to 30 times. While this approach has been proposed as a potential solution to the ocean acidification problem, mitigation of surface ocean acidification might increase acidification in the less-inhabited deep ocean.

A report by the UK's Royal Society (2009) reviewed the approach for effectiveness, affordability, timeliness and safety. The rating for affordability was "medium", or "not expected to be very cost-effective." For the other three criteria, the ratings ranged from "low" to "very low" (i.e., not good). For example, in regards to safety, the report found a "[high] potential for undesirable ecological side effects," and that ocean fertilization "may increase anoxic regions of ocean ('dead zones')."

Carbon Negative Fuels

Carbonic acid can be extracted from seawater as carbon dioxide for use in making synthetic fuel.

If the resulting flue exhaust gas was subject to carbon capture, then the process would be carbon negative over time, resulting in permanent extraction of inorganic carbon from seawater and the atmosphere with which seawater is in equilibrium. Based on the energy requirements, this process was estimated to cost about $50 per tonne of CO_2.

Thermal Pollution

Thermal pollution is the degradation of water quality by any process that changes ambient water temperature. A common cause of thermal pollution is the use of water as a coolant by power plants and industrial manufacturers. When water used as a coolant is returned to the natural environment at a higher temperature, the change in temperature decreases oxygen supply and affects ecosystem composition. Fish and other organisms adapted to particular temperature range can be killed by an abrupt change in water temperature (either a rapid increase or decrease) known as "thermal shock."

The Brayton Point Power Station in Massachusetts discharges heated water to Mount Hope Bay.

Urban runoff—stormwater discharged to surface waters from roads and parking lots—can also be a source of elevated water temperatures.

Ecological Effects

Potrero Generating Station discharged heated water into San Francisco Bay. The plant was closed in 2011.

Warm Water

Elevated temperature typically decreases the level of dissolved oxygen of water. This can harm aquatic animals such as fish, amphibians and other aquatic organisms. Thermal pollution may also increase the metabolic rate of aquatic animals, as enzyme activity, resulting in these organisms consuming more food in a shorter time than if their environment were not changed. An increased metabolic rate may result in fewer resources; the more adapted organisms moving in may have an advantage over organisms that are not used to the warmer temperature. As a result, food chains of the old and new environments may be compromised. Some fish species will avoid stream segments or coastal areas adjacent to a thermal discharge. Biodiversity can be decreased as a result.

High temperature limits oxygen dispersion into deeper waters, contributing to anaerobic conditions. This can lead to increased bacteria levels when there is ample food supply. Many aquatic species will fail to reproduce at elevated temperatures.

Primary producers (e.g. plants, cyanobacteria) are affected by warm water because higher water temperature increases plant growth rates, resulting in a shorter lifespan and species overpopulation. This can cause an algae bloom which reduces oxygen levels.

Temperature changes of even one to two degrees Celsius can cause significant changes in organism metabolism and other adverse cellular biology effects. Principal adverse changes can include rendering cell walls less permeable to necessary osmosis, coagulation of cell proteins, and alteration of enzyme metabolism. These cellular level effects can adversely affect mortality and reproduction.

A large increase in temperature can lead to the denaturing of life-supporting enzymes by breaking down hydrogen- and disulphide bonds within the quaternary structure of the enzymes. Decreased enzyme activity in aquatic organisms can cause problems such as the inability to break down lipids, which leads to malnutrition.

In limited cases, warm water has little deleterious effect and may even lead to improved function of the receiving aquatic ecosystem. This phenomenon is seen especially in seasonal waters and is known as thermal enrichment. An extreme case is derived from the aggregational habits of the manatee, which often uses power plant discharge sites during winter. Projections suggest that manatee populations would decline upon the removal of these discharges.

Cold Water

Releases of unnaturally cold water from reservoirs can dramatically change the fish and macroinvertebrate fauna of rivers, and reduce river productivity. In Australia, where many rivers have warmer temperature regimes, native fish species have been eliminated, and macroinvertebrate fauna have been drastically altered. This may be mitigated by designing the dam to release warmer surface waters instead of the colder water at the bottom of the reservoir.

Thermal Shock

When a power plant first opens or shuts down for repair or other causes, fish and other organisms adapted to particular temperature range can be killed by the abrupt change in water temperature, either an increase or decrease, known as "thermal shock."

Sources and Control of Thermal Pollution

Cooling tower at Gustav Knepper Power Station, Dortmund, Germany

Industrial Wastewater

In the United States, about 75 to 82 percent of thermal pollution is generated by power plants.[335] The remainder is from industrial sources such as petroleum refineries, pulp and paper mills, chemical plants, steel mills and smelters. Heated water from these sources may be controlled with:

- cooling ponds, man-made bodies of water designed for cooling by evaporation, convection, and radiation

- cooling towers, which transfer waste heat to the atmosphere through evaporation and/or heat transfer

- cogeneration, a process where waste heat is recycled for domestic and/or industrial heating purposes.

Some facilities use once-through cooling (OTC) systems which do not reduce temperature as effectively as the above systems. For example, the Potrero Generating Station in San Francisco (closed in 2011), used OTC and discharged water to San Francisco Bay approximately 10 °C (20 °F) above the ambient bay temperature.

A bioretention cell for treating urban runoff in California

Urban Runoff

During warm weather, urban runoff can have significant thermal impacts on small streams, as storm water passes over hot parking lots, roads and sidewalks. Storm water management facilities that absorb runoff or direct it into groundwater, such as bioretention systems and infiltration basins, can reduce these thermal effects. These and related systems for managing runoff are components of an expanding urban design approach commonly called green infrastructure.

Retention basins (stormwater ponds) tend to be less effective at reducing runoff temperature, as the water may be heated by the sun before being discharged to a receiving stream.

Nutrient Pollution

Nutrient pollution caused by Surface runoff of soil and fertilizer during a rain storm

Nutrient pollution, a form of water pollution, refers to contamination by excessive inputs of nutrients. It is a primary cause of eutrophication of surface waters, in which excess nutrients, usually nitrogen or phosphorus, stimulate algal growth. Sources of nutrient pollution include surface runoff from farm fields and pastures, discharges from septic tanks and feedlots, and emissions from combustion. Excess nutrients have been summarized as potentially leading to:

- Population effects: excess growth of algae (blooms);

- Community effects: species composition shifts (dominant taxa);

- Ecological effects:food web changes, light limitation;

- Biogeochemical effects: excess organic carbon (eutrophication); dissolved oxygen deficits (environmental hypoxia); toxin production;

- Human health effects: excess nitrate in drinking water (blue baby syndrome); disinfection by-products in drinking water.

In a 2011 United States Environmental Protection Agency report, the agency's Science Advisory Board succinctly stated: "Excess reactive nitrogen compounds in the environment are associated with many large-scale environmental concerns, including eutrophication of surface waters, toxic algae blooms, hypoxia, acid rain, nitrogen saturation in forests, and global warming."

Excess Nutrients and TMDLs

The regulatory mechanism in the United States, a Total Maximum Daily Load (TMDL), prescribes the maximum amount of a pollutant (including nutrients) that a body of water can receive while still meeting U.S. Clean Water Act water quality standards. Specifically, Section 303 of the Clean Water Act requires each state to generate a TMDL report for each body of water impaired by pollutants. TMDL reports identify pollutant levels and strategies to accomplish pollutant reduction goals. EPA has described TMDLs as establishing a pollutant budget then allocating portions of the overall budget to the pollutant's sources. For many coastal water bodies, the main pollutant issue is excess nutrients, also termed *nutrient over-enrichment*. A TMDL can prescribe the minimum level of Dissolved Oxygen (DO) available in a body of water, which is directly related to nutrient levels. In 2010, 18 percent of TMDLs nationwide were related to nutrient levels including organic enrichment/oxygen depletion, noxious plants, algal growth, and ammo-nia. In Long Island Sound the TMDL development process enabled the Connecticut Department of Environmental Protection (CTDEP) and the New York State Department of Environmental Conservation (NYSDEC) to incorporate a 58.5 percent nitrogen reduction target into a regulatory and legal framework.

Nutrient Remediation

Mussels, nutrient bioextractors.

Innovative solutions have been conceived to deal with nutrient pollution in aquatic systems by altering or enhancing natural processes to shift nutrient effects away from detrimental ecological impacts. Nutrient remediation is a form of environmental remediation, but concerns only biologically active nutrients such as nitrogen and phosphorus. "Remediation" refers to the removal of pollution or contaminants, generally for the protection of human health. In environmental remediation nutrient removal technologies include biofiltration, which uses living material to capture and biologically degrade pollutants. Examples include green belts, riparian areas, natural and constructed wetlands, and treatment ponds. These areas most commonly capture anthropogenic discharges such as wastewater, stormwater runoff, or sewage treatment, for land reclamation after mining, refinery activity, or land development. Biofiltration utilizes biological assimilation to capture, absorb, and eventually incorporate the pollutants (including nutrients) into living tissue. Another form of nutrient removal is bioremediation, which uses microrganisms to remove pollutants. Bioremediation can occur on its own as natural attenuation or intrinsic bioremediation or can be encouraged by the addition of fertilizers, called biostimulation.

Nutrient bioextraction is the preferred term for bioremediation involving cultured plants and animals. Nutrient bioextraction or bioharvesting is the practice of farming and harvesting shellfish and seaweed for the purpose of removing nitrogen and other nutrients from natural water bodies. It has been suggested that nitrogen removal by oyster reefs could generate net benefits for sources facing nitrogen emission restrictions, similar to other nutrient trading scenarios. Specifically, if oysters maintain nitrogen levels in estuaries below thresholds that would lead to the imposition of emission limits, oysters effectively save the sources the compliance costs they otherwise would incur. Several studies have shown that oysters and mussels have the capacity to dramatically impact nitrogen levels in estuaries.

History of Nutrient Policy in the United States

In 1998, a Policy for a National Nutrient Strategy was created with a focus on developing nutrient criteria. Between 2000-2010 criteria for Rivers/Streams, Lakes/Reservoirs, Estuaries, Wetlands, and guidance were completed, including "ecoregional" nutrient criteria in 14 ecoregions across the U.S. In 2004, the EPA Office of Science and Technology defined EPA's expectations for numeric criteria for total nitrogen (TN), total phosphorus (TP), chlorophyll a(chl-a), and clarity, and established "mutually-agreed upon plans" for state criteria development. In 2007, EPA reiterated EPA's expectations for numeric criteria and committed EPA to support state efforts.

Nutrient Trading

After the EPA had introduced watershed-based NPDES permitting in 2007, interest in nutrient removal and achieving regional TMDLs led to the development of nutrient trading schemes. Nutrient trading is a type of water quality trading, a market-based policy instrument used to improve or maintain water quality. Water quality trading arose around 2005 and is based on the fact that different pollution sources in a watershed can face very different costs to control the same pollutant. Water quality trading involves the voluntary exchange of pollution reduction credits from sources with low costs of pollution control to those with high costs of pollution control, and the same principles apply to nutrient water quality trading. The underlying principle is "polluter pays", usually linked with a regulatory driver for participating is the trading program.

A 2013 Forest Trends report summarized water quality trading programs and found three main types of funders: beneficiaries of watershed protection, polluters compensating for their impacts and 'public good payers' that may not directly benefit, but fund the pollution reduction credits on behalf of a government or NGO. As of 2013, payments were overwhelmingly initiated by public good payers like governments and NGOs.

Nonpoint Source Pollution

Muddy river

Non-point source (NPS) pollution includes both water and air pollution from diffuse sources. Non-point source water pollution affects a water body from sources such as polluted runoff from agricultural areas draining into a river, or wind-borne debris blowing out to sea. Non-point source air pollution affects air quality from sources such as smokestacks or car tailpipes. Although these pollutants have originated from a point source, the long-range transport ability and multiple sources of the pollutant make it a non-point source of pollution. Non-point source pollution can be contrasted with point source pollution, where discharges occur to a body of water or into the atmosphere at a single location.

NPS may derive from many different sources with no specific solution may change to rectify the problem, making it difficult to regulate. Non point source water pollution is difficult to control because it comes from the everyday activities of many different people, such as fertilizing a lawn, using a pesticide, or constructing a road or building.

It is the leading cause of water pollution in the United States today, with polluted runoff from agriculture the primary cause.

Other significant sources of runoff include hydrological and habitat modification, and silviculture (forestry).

Contaminated storm-water washed off parking lots, roads and highways, and lawns (often containing fertilizers and pesticides) is called urban runoff. This runoff is often classified as a type of NPS pollution. Some people may also consider it a point source because many times it is channeled into municipal storm drain systems and discharged through pipes to nearby surface waters. However, not all urban runoff flows through storm drain systems before entering water bodies. Some may flow directly into water bodies, especially in developing and suburban areas. Also, unlike other types of point sources, such as industrial discharges, sewage treatment plants and other operations, pollution in urban runoff cannot be attributed to one activity or even group of activities. Therefore, because it is not caused by an easily identified and regulated activity, urban runoff pollution sources are also often treated as true non-point sources as municipalities work to abate them.

Principal Types

Runoff of soil and fertilizer during a rain storm

Sediment

Sediment (loose soil) includes silt (fine particles) and suspended solids (larger particles). Sediment may enter surface waters from eroding stream banks, and from surface runoff due to improper plant cover on urban and rural land. Sediment creates turbidity (cloudiness) in water bodies, reducing the amount of light reaching lower depths, which can inhibit growth of submerged aquatic plants and consequently affect species which are dependent on them, such as fish and shellfish. High turbidity levels also inhibit drinking water purification systems.

Sediment can also be discharged from multiple different sources. Sources include construction sites (although these are point sources, which can be managed with erosion controls and sediment controls), agricultural fields, stream banks, and highly disturbed areas.

Nutrients

Nutrients mainly refers to inorganic matter from runoff, landfills, livestock operations and crop lands. The two primary nutrients of concern are phosphorus and nitrogen.

Phosphorus is a nutrient that occurs in many forms that are bioavailable. It is notoriously over-abundant in human sewage sludge. It is a main ingredient in many fertilizers used for agriculture as well as on residential and commercial properties, and may become a limiting nutrient in freshwater systems and some estuaries. Phosphorus is most often transported to water bodies via soil erosion because many forms of phosphorus tend to be adsorbed on to soil particles. Excess amounts of phosphorus in aquatic systems (particularly freshwater lakes, reservoirs, and ponds) leads to proliferation of microscopic algae called phytoplankton. The increase of organic matter supply due to the excessive growth of the phytoplankton is called eutrophication. A common symptom of eutrophication is algae blooms that can produce unsightly surface scums, shade out beneficial types of plants, and poison the water due to toxins produced by the algae. These toxins are a particular problem in systems used for drinking water because some toxins can cause human illness and removal of the toxins is difficult and expensive. Bacterial decomposition of algal blooms consumes dissolved oxygen in the water, generating hypoxia with detrimental consequences for fish and aquatic invertebrates.

Nitrogen is the other key ingredient in fertilizers, and it generally becomes a pollutant in saltwater or brackish estuarine systems where nitrogen is a limiting nutrient. Similar to phosphorus in fresh-waters, excess amounts of bioavailable nitrogen in marine systems lead to eutrophication and algae blooms. Hypoxia is an increasingly common result of eutrophication in marine systems and can impact large areas of estuaries, bays, and near shore coastal waters. Each summer, hypoxic conditions form in bottom waters where the Mississippi River enters the Gulf of Mexico. During recent summers, the aereal extent of this "dead zone" is comparable to the area of New Jersey and has major detrimental consequences for fisheries in the region.

Nitrogen is most often transported by water as nitrate (NO_3). The nitrogen is usually added to a watershed as organic-N or ammonia (NH_3), so nitrogen stays attached to the soil until oxidation converts it into nitrate. Since the nitrate is generally already incorporated into the soil, the water traveling through the soil (i.e., interflow and tile drainage) is the most likely to transport it, rather than surface runoff.

Toxic Contaminants and Chemicals

Compounds including heavy metals like lead, mercury, zinc, and cadmium, organics like polychlorinated biphenyls (PCBs) and polycyclic aromatic hydrocarbons (PAHs), fire retardants, and other substances are resistant to breakdown. These contaminants can come from a variety of sources including human sewage sludge, mining operations, vehicle emissions, fossil fuel combustion, urban runoff, industrial operations and landfills.

Toxic chemicals mainly include organic compounds and inorganic compounds. These compounds include pesticides like DDT, acids, and salts that have severe effects to the ecosystem and water-bodies. These compounds can threaten the health of both humans and aquatic species while being resistant to environmental breakdown, thus allowing them to persist in the environment. These toxic chemicals could come from croplands, nurseries, orchards, building sites, gardens, lawns and landfills.

Acids and salts mainly are inorganic pollutants from irrigated lands, mining operations, urban runoff, industrial sites and landfills.

Pathogens

Pathogens are bacteria and viruses that can be found in water and cause diseases in humans. Typically, pathogens cause disease when they are present in public drinking water supplies. Pathogens found in contaminated runoff may include:

- *Cryptosporidium parvum*

- *Giardia lamblia*

- *Salmonella*

- *Novovirus* and other viruses

- Parasitic worms (helminths).

Coliform bacteria and fecal matter may also be detected in runoff. These bacteria are a commonly used indicator of water pollution, but not an actual cause of disease.

Pathogens may contaminate runoff due to poorly managed livestock operations, faulty septic systems, improper handling of pet waste, the over application of human sewage sludge, contaminated storm sewers, and sanitary sewer overflows.

Principal Sources

Urban and Suburban Areas

Urban and suburban areas are a main sources of nonpoint source pollution due to the amount of runoff that is produced due to the large amount of paved surfaces. Paved surfaces, such as asphalt and concrete are impervious to water penetrating them. Any water that is on contact with these surfaces will run off and be absorbed by the surrounding environment. These surfaces make it easier for stormwater to carry pollutants into the surrounding soil.

Construction sites tend to have disturbed soil that is easily eroded by precipitation like rain, snow, and hail. Additionally, discarded debris on the site can be carried away by runoff waters and enter the aquatic environment.

Typically, in suburban areas, chemicals are used for lawn care. These chemicals can end up in runoff and enter the surrounding environment via storm drains in the city. Since the water in storm drains is not treated before flowing into surrounding water bodies, the chemicals enter the water directly.

Agricultural Operations

Agricultural operations account for a large percentage of all nonpoint source pollution in the United States. When large tracts of land are plowed to grow crops, it exposes and loosens soil that was once buried. This makes the exposed soil more vulnerable to erosion during rainstorms. It also can increase the amount of fertilizer and pesticides carried into nearby bodies of water.

Atmospheric Inputs

Atmospheric inputs of pollution into the air can come from multiple sources. Typically, industrial facilities, like factories, emit air pollution via a smokestack. Although this is a point source, due to the distributional nature, long-range transport, and multiple sources of the pollution, it is considered a nonpoint source. Additionally, atmospheric pollution can become water pollution, by being washed out of the atmosphere in the form of rain or snow.

Highway Runoff

Highway runoff accounts for a small but widespread percentage of all nonpoint source pollution. Harned (1988) estimated that runoff loads were composed of atmospheric fallout (9%), vehicle deposition (25%) and highway maintenance materials (67%) he also estimated that about 9 percent of these loads were reentrained in the atmosphere.

Forestry and Mining Operations

Forestry and mining operations can have significant inputs to non-point source pollution.

Forestry

Forestry operations reduce the number of trees in a given area, thus reducing the oxygen levels in that area as well. This action, coupled with the heavy machinery rolling over the soil increases the risk of erosion.

Mining

Active mining operations are considered point sources, however runoff from abandoned mining operations contribute to nonpoint source pollution. In strip mining operations, the top of the mountain is removed to expose the desired ore. If this area is not properly reclaimed once the mining has finished, soil erosion can occur. Additionally, there can be chemical reactions with the air and newly exposed rock to create acidic runoff. Water that seeps out of abandoned subsurface mines can also be highly acidic. This can seep into the nearest body of water and change the pH in the aquatic environment.

Marinas and Boating Activities

Chemicals used for boat maintenance, like paint, solvents, and oils find their way into water through runoff. Additionally, spilling fuels or leaking fuels directly into the water from boats contribute to nonpoint source pollution. Nutrient and bacteria levels are increased by poorly maintained sanitary waste receptacles on the boat and pump-out stations.

Control

Contour buffer strips used to retain soil and reduce erosion

Urban and Suburban Areas

To control non-point source pollution, many different approaches can be undertaken in both urban and suburban areas. Buffer strips provide a barrier of grass in between impervious paving material like parking lots and roads, and the closest body of water. This allows the soil to absorb any pollution before it enters the local aquatic system. Retention ponds can be built in drainage areas to create an aquatic buffer between runoff pollution and the aquatic environment. Runoff and storm water drain into the retention pond allowing for the contaminates to settle out and become trapped in the pond. The use of porous pavement allows for rain and storm water to drain into the ground beneath the pavement, reducing the amount of runoff that drains directly into the water body. Restoration methods such as constructing wetlands are also used to slow runoff as well as absorb contamination.

Construction sites typically implement simple measures to reduce pollution and runoff. Firstly, sediment or silt fences are erected around construction sites to reduce the amount of sediment and large material draining into the nearby water body. Secondly, laying grass or straw along the border of construction sites also work to reduce nonpoint source pollution.

In areas served by single-home septic systems, local government regulations can force septic system maintenance to ensure compliance with water quality standards. In Washington (state), a novel approach to was developed through creation of a "shellfish protection district" when either a commercial or recreational shellfish bed is downgraded because of ongoing nonpoint source pollution. The shellfish protection district is a geographic area designated by a county to protect water quality and tideland resources, and provides a mechanism to generate local funds for water quality services to control nonpoint sources of pollution. At least two shellfish protection districts in south Puget Sound have instituted septic system operation and maintenance requirements with program fees tied directly to property taxes.

Agricultural Operations

To control sediment and runoff, farmers may utilize erosion controls to reduce runoff flows and retain soil on their fields. Common techniques include contour plowing, crop mulching, crop rotation, planting perennial crops and installing riparian buffers. *Conservation tillage* is a concept used to reduce runoff while planting a new crop. The farmer leaves some crop reside from the previous planting in the ground to help prevent runoff during the planting process.

Nutrients are typically applied to farmland as commercial fertilizer; animal manure; or spraying of municipal or industrial wastewater (effluent) or sludge. Nutrients may also enter runoff from crop residues, irrigation water, wildlife, and atmospheric deposition. Farmers can develop and implement nutrient management plans to reduce excess application of nutrients.

To minimize pesticide impacts, farmers may use Integrated Pest Management (IPM) techniques (which can include biological pest control) to maintain control over pests, reduce reliance on chemical pesticides, and protect water quality.

Forestry Operations

With well planned placement of both logging trails, also called skid trails, can reduce the amount of sediment generated. By planning the trails location as far away from the logging activity as possible as well as contouring the trails with the land, it can reduce the amount of loose sediment in the runoff. Additionally, by replanting trees on the land after logging, it provides a structure for the soil to regain stability as well as replaces the logged environment.

Marinas

Installing shut off valves on fuel pumps at a marina dock can help reduce the amount of spillover into the water. Additionally, pump-out stations that are easily accessible to boaters in a marina can provide a clean place in which to dispose of sanitary waste without dumping it directly into the water. Finally, something as simple as having trash containers around a marina can prevent larger objects entering the water.

References

- For example, see Baird, Rodger B.; Clesceri, Leonore S.; Eaton, Andrew D.; et al., eds. (2012). Standard Methods for the Examination of Water and Wastewater (22nd ed.). Washington, DC: American Public Health Association. ISBN 978-0875530130.

- Kennish, Michael J. (1992). Ecology of Estuaries: Anthropogenic Effects. Marine Science Series. Boca Raton, FL: CRC Press. pp. 415–17. ISBN 978-0-8493-8041-9.

- G. Allen Burton, Jr., Robert Pitt (2001). Stormwater Effects Handbook: A Toolbox for Watershed Managers, Scientists, and Engineers. New York: CRC/Lewis Publishers. ISBN 0-87371-924-7. Chapter 2.

- North, W. J.; James, D. E.; Jones, L. G. (1993). "History of kelp beds (Macrocystis) in Orange and San Diego Counties, California". Fourteenth International Seaweed Symposium. p. 277. doi:10.1007/978-94-011-1998-6_33. ISBN 978-94-010-4882-8.

- Kump, Lee R.; Kasting, James F.; Crane, Robert G. (2003). The Earth System (2nd ed.). Upper Saddle River: Prentice Hall. pp. 162–164. ISBN 0-613-91814-2.

- Kennish, Michael J. (1992). Ecology of Estuaries: Anthropogenic Effects. Marine Science Series. Boca Raton, FL: CRC Press. ISBN 978-0-8493-8041-9.

- Kump, L.R.; Bralower, T.J.; Ridgwell, A. (2009). "Ocean acidification in deep time". Oceanography. 22: 94–107. doi:10.5670/oceanog.2009.10. Retrieved 16 May 2016.

- Genevieve Bennett, Nathaniel Carroll, and Katherine Hamilton (2013). "Charting New Waters State of Watershed Payments 2012Figure 4b: Watershed Investments by Payer Type, Globally, Excluding China" (PDF). Forest Trends Association. p. 98. Retrieved 20 February 2015.

- Administration, US Department of Commerce, National Oceanic and Atmospheric. "What is the biggest source of pollution in the ocean?". oceanservice.noaa.gov. Retrieved 2015-11-22.

- "Brayton Point Station: Final NPDES Permit". NPDES Permits in New England. U.S. Environmental Protection Agency (EPA), Boston, MA. 2014. Retrieved 2015-04-13.

- "Total Maximum Daily Load (TMDLs) at Work: New York: Restoring the Long Island Sound While Saving Money". EPA. Retrieved June 14, 2013.

- "Total Maximum Daily Load (TMDLs) at Work: New York: Restoring the Long Island Sound While Saving Money". EPA. Retrieved June 14, 2013.

5

Approaches and Strategies of Water Conservation

This chapter discusses the strategies of water conservation in a critical manner providing key analysis to the subject matter. It essentially explains to the reader approaches such as water resource management, wastewater treatment, rainwater harvesting and water metering. The aspects elucidated in this chapter are of vital importance, and provide a better understanding of water conservation.

Water Resource Management

Water resource management is the activity of planning, developing, distributing and managing the optimum use of water resources. It is a sub-set of water cycle management. Ideally, water resource management planning has regard to all the competing demands for water and seeks to allocate water on an equitable basis to satisfy all uses and demands. As with other resource management, this is rarely possible in practice.

Overview

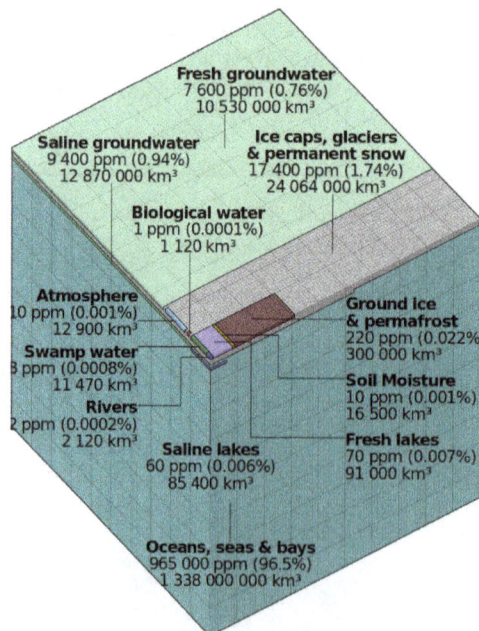

Visualisation of the distribution (by volume) of water on Earth. Each tiny cube (such as the one representing biological water) corresponds to approximately 1000 cubic km of water, with a mass of approximately 1 trillion tonnes (2000 times that of the Great Pyramid of Giza or 5 times that of Lake Kariba, arguably the heaviest man-made object). The entire block comprises 1 million tiny cubes.

Water is an essential resource for all life on the planet. Of the water resources on Earth only three percent of it is fresh and two-thirds of the freshwater is locked up in ice caps and glaciers. Of the remaining one percent, a fifth is in remote, inaccessible areas and much seasonal rainfall in monsoonal deluges and floods cannot easily be used. As time advances, water is becoming scarcer and having access to clean, safe, drinking water is limited among countries. At present only about 0.08 percent of all the world's fresh water is exploited by mankind in ever increasing demand for sanitation, drinking, manufacturing, leisure and agriculture. Due to the small percentage of water remaining, optimizing the fresh water we have left from natural resources has been a continuous difficulty in several locations worldwide.

Much efforts in water resource management is directed at optimizing the use of water and in minimizing the environmental impact of water use on the natural environment. The observation of water as an integral part of the ecosystem is based on integrated water resource management, where the quantity and quality of the ecosystem help to determine the nature of the natural resources.

Successful management of any resources requires accurate knowledge of the resource available, the uses to which it may be put, the competing demands for the resource, measures to and processes to evaluate the significance and worth of competing demands and mechanisms to translate policy decisions into actions on the ground.

For water as a resource this is particularly difficult since sources of water can cross many national boundaries and the uses of water include many that are difficult to assign financial value to and may also be difficult to manage in conventional terms. Examples include rare species or ecosystems or the very long term value of ancient ground water reserves.

Agriculture

Agriculture is the largest user of the world's freshwater resources, consuming 70 percent. As the world population rises it consumes more food (currently exceeding 6%, it is expected to reach 9% by 2050), the industries and urban developments expand, and the emerging biofuel crops trade also demands a share of freshwater resources, water scarcity is becoming an important issue. An assessment of water resource management in agriculture was conducted in 2007 by the International Water Management Institute in Sri Lanka to see if the world had sufficient water to provide food for its growing population or not . It assessed the current availability of water for agriculture on a global scale and mapped out locations suffering from water scarcity. It found that a fifth of the world's people, more than 1.2 billion, live in areas of physical water scarcity, where there is not enough water to meet all their demands. A further 1.6 billion people live in areas experiencing economic water scarcity, where the lack of investment in water or insufficient human capacity make it impossible for authorities to satisfy the demand for water.

The report found that it would be possible to produce the food required in future, but that continuation of today's food production and environmental trends would lead to crises in many parts of the world. Regarding food production, the World Bank targets agricultural food production and water resource management as an increasingly global issue that is fostering an important and growing debate. The authors of the book *Out of Water: From abundance to Scarcity and How to Solve the World's Water Problems*, which laid down a six-point plan for solving the world's water problems. These are: 1) Improve data related to water; 2) Treasure the environment; 3) Reform water governance; 4) Revitalize agricultural water use; 5) Manage urban and industrial demand; and 6) Empower the poor and women in water resource management. To avoid a global water crisis, farmers will have to strive to increase productivity to meet growing demands for food, while industry and cities find ways to use water more efficiently.

Managing Water in Urban Settings

As the carrying capacity of the Earth increases greatly due to technological advances, urbanization in modern times occurs because of economic opportunity. This rapid urbanization happens worldwide but mostly in new rising economies and developing countries. Cities in Africa and Asia are growing fastest with 28 out of 39 megacities (a city or urban area with more than 10 million inhabitants) worldwide in these developing nations. The number of megacities will continue to rise reaching approximately 50 in 2025. With developing economies water scarcity is a very common and very prevalent issue. Global freshwater resources dwindle in the eastern hemisphere either than at the poles, and with the majority of urban development millions live with insufficient fresh water. This is caused by polluted freshwater resources, overexploited groundwater resources, insufficient harvesting capacities in the surrounding rural areas, poorly constructed and maintained water supply systems, high amount of informal water use and insufficient technical and water management capacities.

In the areas surrounding urban centres, agriculture must compete with industry and municipal users for safe water supplies, while traditional water sources are becoming polluted with urban wastewater. As cities offer the best opportunities for selling produce, farmers often have no alternative to using polluted water to irrigate their crops. Depending on how developed a city's wastewater treatment is, there can be significant health hazards related to the use of this water. Wastewater from cities can contain a mixture of pollutants. There is usually wastewater from kitchens and toilets along with rainwater runoff. This means that the water usually contains excessive levels of nutrients and salts, as well as a wide range of pathogens. Heavy metals may also be present, along with traces of antibiotics and endocrine disruptors, such as oestrogens.

Developing world countries tend to have the lowest levels of wastewater treatment. Often, the water that farmers use for irrigating crops is contaminated with pathogens from sewage. The pathogens of most concern are bacteria, viruses and parasitic worms, which directly affect farmers' health and indirectly affect consumers if they eat the contaminated crops. Common illnesses include diarrhoea, which kills 1.1 million people annually and is the second most common cause of infant deaths. Many cholera outbreaks are also related to the reuse of poorly treated wastewater. Actions that reduce or remove contamination, therefore, have the potential to save a large number of lives and improve livelihoods. Scientists have been working to find ways to reduce contamination of food using a method called the 'multiple-barrier approach'.

This involves analysing the food production process from growing crops to selling them in markets and eating them, then considering where it might be possible to create a barrier against contamination. Barriers include: introducing safer irrigation practices; promoting on-farm wastewater treatment; taking actions that cause pathogens to die off; and effectively washing crops after harvest in markets and restaurants.

Future of Water Resources

One of the biggest concerns for our water-based resources in the future is the sustainability of the current and even future water resource allocation. As water becomes more scarce, the importance of how it is managed grows vastly. Finding a balance between what is needed by humans and what is needed in the environment is an important step in the sustainability of water resources. Attempts to create sustainable freshwater systems have been seen on a national level in countries such as Australia, and such commitment to the environment could set a model for the rest of the world.

The field of water resources management will have to continue to adapt to the current and future issues facing the allocation of water. With the growing uncertainties of global climate change and the long term impacts of management actions,the decision-making will be even more difficult. It is likely that ongoing climate change will lead to situations that have not been encountered. As a result, new management strategies will have to be implemented in order to avoid setbacks in the allocation of water resources.

Wastewater Treatment

Wastewater treatment is a process used to convert wastewater - which is water no longer needed or suitable for its most recent use - into an effluent that can be either returned to the water cycle with minimal environmental issues or reused. The latter is called water reclamation and implies avoidance of disposal by use of treated wastewater effluent for various purposes. Treatment means removing impurities from water being treated; and some methods of treatment are applicable to both water and wastewater. The physical infrastructure used for wastewater treatment is called a "wastewater treatment plant" (WWTP).

Bundesarchiv, Bild 183-1984-1002-002
Foto: Pätzold, Ralf | 2. Oktober 1984

Clarifiers are widely used for wastewater treatment.

The treatment of wastewater belongs to the overarching field of Public Works - Environmental, with the management of human waste, solid waste, sewage treatment, stormwater (drainage) management, and water treatment. By-products from wastewater treatment plants, such as screenings, grit and sewage sludge may also be treated in a wastewater treatment plant. If the wastewater is predominantly from municipal sources (households and small industries) it is called sewage and its treatment is called sewage treatment.,

Disposal or Reuse

Although disposal or reuse occurs after treatment, it must be considered first. Since disposal or reuse are the objectives of wastewater treatment, disposal or reuse options are the basis for treatment decisions. Acceptable impurity concentrations may vary with the type of use or location of disposal. Transportation costs often make acceptable impurity concentrations dependent upon location of disposal, but expensive treatment requirements may encourage selection of a disposal location on the basis of impurity concentrations. Ocean disposal is subject to international treaty requirements. International treaties may also regulate disposal into rivers crossing international borders. Water bodies entirely within the jurisdiction of a single nation may be subject to regulations of multiple local governments. Acceptable impurity concentrations may vary widely among different jurisdictions for disposal of wastewater to evaporation ponds, infiltration basins, or injection wells.

Processes Used

Phase Separation

Phase separation transfers impurities into a non-aqueous phase. Phase separation may occur at intermediate points in a treatment sequence to remove solids generated during oxidation or polishing. Grease and oil may be recovered for fuel or saponification. Solids often require dewatering of sludge in a wastewater treatment plant. Disposal options for dried solids vary with the type and concentration of impurities removed from water.

Production of waste brine, however, may discourage wastewater treatment removing dissolved inorganic solids from water by methods like ion exchange, reverse osmosis, and distillation.

Primary settling tank of wastewater treatment plant in Dresden-Kaditz, Germany

Sedimentation

Solids and non-polar liquids may be removed from wastewater by gravity when density differences are sufficient to overcome dispersion by turbulence. Gravity separation of solids is the primary treatment of sewage, where the unit process is called "primary settling tanks" or "primary sedimentation tanks". It is also widely used for the treatment of other wastewaters. Solids that are heavier than water will accumulate at the bottom of quiescent settling basins. More complex clarifiers also have skimmers to simultaneously remove floating grease like soap scum and solids like feathers or wood chips. Containers like the API oil-water separator are specifically designed to separate non-polar liquids.

Filtration

Colloidal suspensions of fine solids may be removed by filtration through fine physical barriers distinguished from coarser screens or sieves by the ability to remove particles smaller than the openings through which the water passes. Other types of water filters remove impurities by chemical or biological processes described below.

Oxidation

Oxidation reduces the biochemical oxygen demand of wastewater, and may reduce the toxicity of some impurities. Secondary treatment converts some impurities to carbon dioxide, water, and biosolids. Chemical oxidation is widely used for disinfection.

Aeration tank of an activated sludge process at the wastewater treatment plant in Dresden-Kaditz, Germany

Biochemical Oxidation

Secondary treatment by biochemical oxidation of dissolved and colloidal organic compounds is widely used in sewage treatment and is applicable to some agricultural and industrial wastewaters. Biological oxidation will preferentially remove organic compounds useful as a food supply for the treatment ecosystem. Concentration of some less digestable compounds may be reduced by cometabolism. Removal efficiency is limited by the minimum food concentration required to sustain the treatment ecosystem. Removal efficiencies of over 96% are readily attainable through proper process controls and effective maintenance of plant equipment.

Chemical Oxidation

Chemical oxidation may remove some persistent organic pollutants and concentrations remaining after biochemical oxidation. Disinfection by chemical oxidation kills bacteria and microbial pathogens by adding ozone, chlorine or hypochlorite to wastewater.

Polishing

Polishing refers to treatments made following the above methods. These treatments may also be used independently for some industrial wastewater. Chemical reduction or pH adjustment minimizes chemical reactivity of wastewater following chemical oxidation. Carbon filtering removes remaining contaminants and impurities by chemical absorption onto activated carbon. Filtration through sand (calcium carbonate) or fabric filters is the most common method used in municipal wastewater treatment.

Wastewater Treatment Plants

Wastewater treatment plants may be distinguished by the type of wastewater to be treated, i.e. whether it is sewage, industrial wastewater, agricultural wastewater or leachate.

Overview of the wastewater treatment plant of Antwerpen-Zuid,
located in the south of the agglomeration of Antwerp (Belgium)

Sewage Treatment Plants

A typical municipal sewage treatment plant in an industrialized country may include primary treatment to remove solid material, secondary treatment to digest dissolved and suspended organic material as well as the nutrients nitrogen and phosphorus, and - sometimes but not always - disinfection to kill pathogenic bacteria. The sewage sludge that is produced in sewage treatment plants undergoes sludge treatment. Larger municipalities often include factories discharging industrial wastewater into the municipal sewer system. The term "sewage treatment plant" is nowadays often replaced with the term "wastewater treatment plant".

Tertiary Treatment

Tertiary treatment is a term applied to polishing methods used following a traditional sewage treatment sequence. Tertiary treatment is being increasingly applied in industrialized countries and most common technologies are micro filtration or synthetic membranes. After membrane filtration, the treated wastewater is nearly indistinguishable from waters of natural origin of drinking quality (without its minerals). Nitrates can be removed from wastewater by natural processes in wetlands but also via microbial denitrification. Ozone wastewater treatment is also growing in popularity, and requires the use of an ozone generator, which decontaminates the water as ozone bubbles percolate through the tank, but this treatment is energy intensive. Latest, and very promising treatment technology is the use aerobic granulation.

Industrial Wastewater Treatment Plants

Disposal of wastewaters from an industrial plant is a difficult and costly problem. Most petroleum refineries, chemical and petrochemical plants have onsite facilities to treat their wastewaters so that the pollutant concentrations in the treated wastewater comply with the local and/or national regulations regarding disposal of wastewaters into community treatment plants or into rivers, lakes or oceans. Constructed wetlands are being used in an increasing number of cases as they provided high quality and productive on-site treatment. Other industrial processes that produce a lot of waste-waters such as paper and pulp production has created environmental concern, leading to development of processes to recycle water use within plants before they have to be cleaned and disposed.

Industrial wastewater treatment plants are required where municipal sewage treatment plants are unavailable or cannot adequately treat specific industrial wastewaters. Industrial wastewater plants may reduce raw water costs by converting selected wastewaters to reclaimed water used for different purposes. Industrial wastewater treatment plants may reduce wastewater treatment charges collected by municipal sewage treatment plants by pre-treating wastewaters to reduce concentrations of pollutants measured to determine user fees.

Although economies of scale may favor use of a large municipal sewage treatment plant for disposal of small volumes of industrial wastewater, industrial wastewater treatment and disposal may be less expensive than correctly apportioned costs for larger volumes of industrial wastewater not requiring the conventional sewage treatment sequence of a small municipal sewage treatment plant.

An industrial wastewater treatment plant may include one or more of the following rather than the conventional primary, secondary, and disinfection sequence of sewage treatment:

- An API oil-water separator, for removing separate phase oil from wastewater.

- A clarifier, for removing solids from wastewater.

- A roughing filter, to reduce the biochemical oxygen demand of wastewater.

- A carbon filtration plant, to remove toxic dissolved organic compounds from wastewater.

- An advanced electrodialysis reversal (EDR) system with ion exchange membranes.

Agricultural Wastewater Treatment Plants

Agricultural wastewater treatment for continuous confined animal operations like milk and egg production may be performed in plants using mechanized treatment units similar to those described under industrial wastewater; but where land is available for ponds, settling basins and facultative lagoons may have lower operational costs for seasonal use conditions from breeding or harvest cycles.

Leachate Treatment Plants

Leachate treatment plants are used to treat leachate from landfills. Treatment options include: biological treatment, mechanical treatment by ultrafiltration, treatment with active carbon filters and reverse osmosis using disc tube module technology.

Subfield of Wastewater Treatment

Industrial Wastewater Treatment

Industrial wastewater treatment covers the mechanisms and processes used to treat wastewater that is produced as a by-product of industrial or commercial activities. After treatment, the treated industrial wastewater (or effluent) may be reused or released to a sanitary sewer or to a surface water in the environment. Most industries produce some wastewater although recent trends in the developed world have been to minimise such production or recycle such wastewater within the production process. However, many industries remain dependent on processes that produce wastewaters.

Sources of Industrial Wastewater

Complex Organic Chemicals Industry

A range of industries manufacture or use complex organic chemicals. These include pesticides, pharmaceuticals, paints and dyes, petrochemicals, detergents, plastics, paper pollution, etc. Waste waters can be contaminated by feedstock materials, by-products, product material in soluble or particulate form, washing and cleaning agents, solvents and added value products such as plasticisers. Treatment facilities that do not need control of their effluent typically opt for a type of aerobic treatment, i.e. aerated lagoons.

Electric Power Plants

Fossil-fuel power stations, particularly coal-fired plants, are a major source of industrial wastewater. Many of these plants discharge wastewater with significant levels of metals such as lead, mercury, cadmium and chromium, as well as arsenic, selenium and nitrogen compounds (nitrates and nitrites). Wastewater streams include flue-gas desulfurization, fly ash, bottom ash and flue gas mercury control. Plants with air pollution controls such as wet scrubbers typically transfer the captured pollutants to the wastewater stream.

Ash ponds, a type of surface impoundment, are a widely used treatment technology at coal-fired plants. These ponds use gravity to settle out large particulates (measured as total suspended solids) from power plant wastewater. This technology does not treat dissolved pollutants. Power stations use additional technologies to control pollutants, depending on the particular wastestream in the plant. These include dry ash handling, closed-loop ash recycling, chemical precipitation, biological treatment (such as an activated sludge process), and evaporation.

Food Industry

Wastewater generated from agricultural and food operations has distinctive characteristics that set it apart from common municipal wastewater managed by public or private sewage treatment plants throughout the world: it is biodegradable and non-toxic, but has high concentrations of biochemical oxygen demand (BOD) and suspended solids (SS). The constituents of food and agriculture wastewater are often complex to predict, due to the differences in BOD and pH in effluents from vegetable, fruit, and meat products and due to the seasonal nature of food processing and post-harvesting.

Processing of food from raw materials requires large volumes of high grade water. Vegetable washing generates waters with high loads of particulate matter and some dissolved organic matter. It may also contain surfactants.

Animal slaughter and processing produces very strong organic waste from body fluids, such as blood, and gut contents. This wastewater is frequently contaminated by significant levels of antibiotics and growth hormones from the animals and by a variety of pesticides used to control external parasites.

Processing food for sale produces wastes generated from cooking which are often rich in plant organic material and may also contain salt, flavourings, colouring material and acids or alkali. Very significant quantities of oil or fats may also be present.

Iron and Steel Industry

The production of iron from its ores involves powerful reduction reactions in blast furnaces. Cooling waters are inevitably contaminated with products especially ammonia and cyanide. Production of coke from coal in coking plants also requires water cooling and the use of water in by-products separation. Contamination of waste streams includes gasification products such as benzene, naphthalene, anthracene, cyanide, ammonia, phenols, cresols together with a range of more complex organic compounds known collectively as polycyclic aromatic hydrocarbons (PAH).

The conversion of iron or steel into sheet, wire or rods requires hot and cold mechanical transformation stages frequently employing water as a lubricant and coolant. Contaminants include hydraulic oils, tallow and particulate solids. Final treatment of iron and steel products before onward sale into manufacturing includes *pickling* in strong mineral acid to remove rust and prepare the surface for tin or chromium plating or for other surface treatments such as galvanisation or painting. The two acids commonly used are hydrochloric acid and sulfuric acid. Wastewaters include acidic rinse waters together with waste acid. Although many plants operate acid recovery plants (particularly those using hydrochloric acid), where the mineral acid is boiled away from the iron salts, there remains a large volume of highly acid ferrous sulfate or ferrous chloride to be disposed of. Many steel industry wastewaters are contaminated by hydraulic oil, also known as *soluble oil*.

Mines and Quarries

The principal waste-waters associated with mines and quarries are slurries of rock particles in water. These arise from rainfall washing exposed surfaces and haul roads and also from rock washing and grading processes. Volumes of water can be very high, especially rainfall related arisings on large sites. Some specialized separation operations, such as coal washing to separate coal from native rock using density gradients, can produce wastewater contaminated by fine particulate haematite and surfactants. Oils and hydraulic oils are also common contaminants.

Mine wastewater effluent in Peru, with neutralized pH from tailing runoff.

Wastewater from metal mines and ore recovery plants are inevitably contaminated by the minerals present in the native rock formations. Following crushing and extraction of the desirable materials, undesirable materials may enter the wastewater stream. For metal mines, this can include unwanted metals such as zinc and other materials such as arsenic. Extraction of high value metals such as gold and silver may generate slimes containing very fine particles in where physical removal of contaminants becomes particularly difficult.

Additionally, the geologic formations that harbour economically valuable metals such as copper and gold very often consist of sulphide-type ores. The processing entails grinding the rock into fine particles and then extracting the desired metal(s), with the leftover rock being known as tailings. These tailings contain a combination of not only undesirable leftover metals, but also sulphide components which eventually form sulphuric acid upon the exposure to air and water that inevitably occurs when the tailings are disposed of in large impoundments. The resulting acid mine drainage, which is often rich in heavy metals (because acids dissolve metals), is one of the many environmental impacts of mining.

Nuclear Industry

The waste production from the nuclear and radio-chemicals industry is dealt with as *Radioactive waste*.

Pulp and Paper Industry

Effluent from the pulp and paper industry is generally high in suspended solids and BOD. Plants that bleach wood pulp for paper making may generate chloroform, dioxins (including 2,3,7,8-TCDD), furans, phenols and chemical oxygen demand (COD). Stand-alone paper mills using imported pulp may only require simple primary treatment, such as sedimentation or dissolved air flotation. Increased BOD or COD loadings, as well as organic pollutants, may require biological treatment such as activated sludge or upflow anaerobic sludge blanket reactors. For mills with high inorganic loadings like salt, tertiary treatments may be required, either general membrane treatments like ultrafiltration or reverse osmosis or treatments to remove specific contaminants, such as nutrients.

Industrial Oil Contamination

Industrial applications where oil enters the wastewater stream may include vehicle wash bays, workshops, fuel storage depots, transport hubs and power generation. Often the wastewater is discharged into local sewer or trade waste systems and must meet local environmental specifications. Typical contaminants can include solvents, detergents, grit. lubricants and hydrocarbons.

Water Treatment

Many industries have a need to treat water to obtain very high quality water for demanding purposes such as environmental discharge compliance. Water treatment produces organic and mineral sludges from filtration and sedimentation. Ion exchange using natural or synthetic resins removes calcium, magnesium and carbonate ions from water, typically replacing them with sodium, chloride, hydroxyl and/or other ions. Regeneration of ion exchange columns with strong acids and alkalis produces a wastewater rich in hardness ions which are readily precipitated out, especially when in admixture with other wastewater constituents.

Wool Processing

Insecticide residues in fleeces are a particular problem in treating waters generated in wool processing. Animal fats may be present in the wastewater, which if not contaminated, can be recovered for the production of tallow or further rendering.

Treatment of Industrial Wastewater

The various types of contamination of wastewater require a variety of strategies to remove the contamination.

Brine Treatment

Brine treatment involves removing dissolved salt ions from the waste stream. Although similarities to seawater or brackish water desalination exist, industrial brine treatment may contain unique combinations of dissolved ions, such as hardness ions or other metals, necessitating specific processes and equipment.

Brine treatment systems are typically optimized to either reduce the volume of the final discharge for more economic disposal (as disposal costs are often based on volume) or maximize the recovery of fresh water or salts. Brine treatment systems may also be optimized to reduce electricity consumption, chemical usage, or physical footprint.

Brine treatment is commonly encountered when treating cooling tower blowdown, produced water from steam assisted gravity drainage (SAGD), produced water from natural gas extraction such as coal seam gas, frac flowback water, acid mine or acid rock drainage, reverse osmosis reject, chlor-alkali wastewater, pulp and paper mill effluent, and waste streams from food and beverage processing.

Brine treatment technologies may include: membrane filtration processes, such as reverse osmosis; ion exchange processes such as electrodialysis or weak acid cation exchange; or evaporation processes, such as brine concentrators and crystallizers employing mechanical vapour recompression and steam.

Reverse osmosis may not be viable for brine treatment, due to the potential for fouling caused by hardness salts or organic contaminants, or damage to the reverse osmosis membranes from hydrocarbons.

Evaporation processes are the most widespread for brine treatment as they enable the highest degree of concentration, as high as solid salt. They also produce the highest purity effluent, even distillate-quality. Evaporation processes are also more tolerant of organics, hydrocarbons, or hardness salts. However, energy consumption is high and corrosion may be an issue as the prime mover is concentrated salt water. As a result, evaporation systems typically employ titanium or duplex stainless steel materials.

Brine Management

Brine management examines the broader context of brine treatment and may include consideration of government policy and regulations, corporate sustainability, environmental impact, recycling, handling and transport, containment, centralized compared to on-site treatment, avoidance and reduction, technologies, and economics. Brine management shares some issues with leachate management and more general waste management.

Solids Removal

Most solids can be removed using simple sedimentation techniques with the solids recovered as slurry or sludge. Very fine solids and solids with densities close to the density of water pose special problems. In such case filtration or ultrafiltration may be required. Although, flocculation may be used, using alum salts or the addition of polyelectrolytes.

Oils and Grease Removal

The effective removal of oils and grease is dependent on the characteristics of the oil in terms of its suspension state and droplet size, which will in turn affect the choice of separator technology.

Oil pollution in water usually comes in four states, often in combination:

- free oil - large oil droplets sitting on the surface;
- heavy oil, which sits at the bottom, often adhering to solids like dirt;
- emulsified, where the oil droplets are heavily "chopped"; and
- dissolved oil, where the droplets are fully dispersed and not visible. Emulsified oil droplets are the most common in industrial oily wastewater and are extremely difficult to separate.

The methodology for separating the oil is dependent on the oil droplet size. Larger oil droplets such as those in free oil pollution are easily removed, but as the droplets become smaller, some separator technologies perform better than others.

Most separator technologies will have an optimum range of oil droplet sizes that can be effectively treated. This is known as the "micron rating."

Analysing the oily water to determine droplet size can be performed with a video particle analyser. Alternatively, there are commonalities in industries for oil droplet sizes. Larger droplets–greater than 60 microns–are often present in wastewater in workshops, re-fuel areas and depots. Twenty to 50 micron oil droplets often are present in vehicle wash bays, meat processing and dairy manufacturing effluent and aluminium billet cooling towers. Smaller droplets in the range of 10 to 20 microns tend to occur in workshops and condensates.

Each separator technology will have its' own performance curve outlining optimum performance based on oil droplet size. the most common separators are gravity tanks or pits, API oil-water separators or plate packs, chemical treatment via DAFs, centrifuges, media filters and hydrocyclones.

API Separators

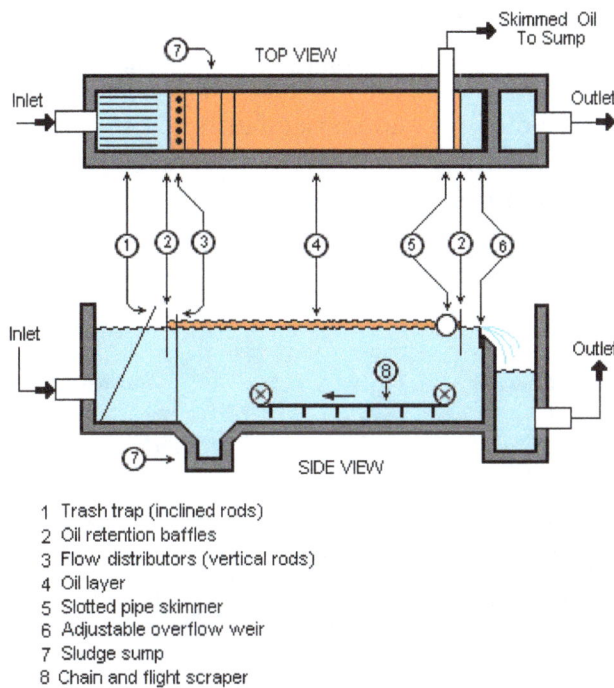

1 Trash trap (inclined rods)
2 Oil retention baffles
3 Flow distributors (vertical rods)
4 Oil layer
5 Slotted pipe skimmer
6 Adjustable overflow weir
7 Sludge sump
8 Chain and flight scraper

A typical API oil-water separator used in many industries

Many oils can be recovered from open water surfaces by skimming devices. Considered a dependable and cheap way to remove oil, grease and other hydrocarbons from water, oil skimmers can sometimes achieve the desired level of water purity. At other times, skimming is also a cost-efficient method to remove most of the oil before using membrane filters and chemical processes. Skimmers will prevent filters from blinding prematurely and keep chemical costs down because there is less oil to process.

Because grease skimming involves higher viscosity hydrocarbons, skimmers must be equipped with heaters powerful enough to keep grease fluid for discharge. If floating grease forms into solid clumps or mats, a spray bar, aerator or mechanical apparatus can be used to facilitate removal.

However, hydraulic oils and the majority of oils that have degraded to any extent will also have a soluble or emulsified component that will require further treatment to eliminate. Dissolving or emulsifying oil using surfactants or solvents usually exacerbates the problem rather than solving it, producing wastewater that is more difficult to treat.

The wastewaters from large-scale industries such as oil refineries, petrochemical plants, chemical plants, and natural gas processing plants commonly contain gross amounts of oil and suspended solids. Those industries use a device known as an API oil-water separator which is designed to separate the oil and suspended solids from their wastewater effluents. The name is derived from the fact that such separators are designed according to standards published by the American Petroleum Institute (API).

The API separator is a gravity separation device designed by using Stokes Law to define the rise velocity of oil droplets based on their density and size. The design is based on the specific gravity difference between the oil and the wastewater because that difference is much smaller than the specific gravity difference between the suspended solids and water. The suspended solids settles to the bottom of the separator as a sediment layer, the oil rises to top of the separator and the cleansed wastewater is the middle layer between the oil layer and the solids.

Typically, the oil layer is skimmed off and subsequently re-processed or disposed of, and the bottom sediment layer is removed by a chain and flight scraper (or similar device) and a sludge pump. The water layer is sent to further treatment for additional removal of any residual oil and then to some type of biological treatment unit for removal of undesirable dissolved chemical compounds.

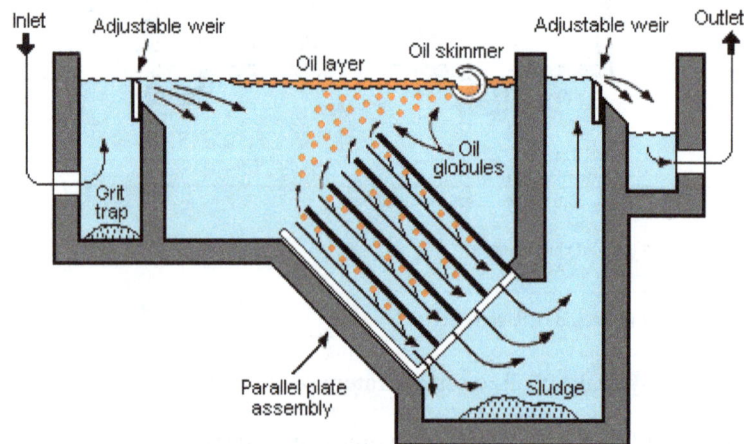

A typical parallel plate separator

Parallel plate separators are similar to API separators but they include tilted parallel plate assemblies (also known as parallel packs). The parallel plates provide more surface for suspended oil droplets to coalesce into larger globules. Such separators still depend upon the specific gravity between the suspended oil and the water. However, the parallel plates enhance the degree of oil-water separation. The result is that a parallel plate separator requires significantly less space than a conventional API separator to achieve the same degree of separation.

Hydrocyclone Oil Separators

Hydrocyclone oil separators operate on the process where wastewater enters the cyclone chamber and is spun under extreme centrifugal forces more than 1000 times the force of gravity. This force causes the water and oil droplets to separate. The separated oil is discharged from one end of the cyclone where treated water is discharged through the opposite end for further treatment, filtration or discharge.

Hydrocyclones are useful for the greatest range of oil droplet sizes operating from less than 10 microns and up and can operate continuously without water pre-treatment and at any temperature and pH. Applications where hydrocyclones are found are in industry where oily water sources arise in workshops, vehicle wash bays, transport hubs, fuel depots and aluminium billet processing. Animal fats from meat processing and dairy manufacturing can also be removed without the need of chemical treatment that often is required for dissolved air flotation (DAF) systems.

Removal of Biodegradable Organics

Biodegradable organic material of plant or animal origin is usually possible to treat using extended conventional sewage treatment processes such as activated sludge or trickling filter. Problems can arise if the wastewater is excessively diluted with washing water or is highly concentrated such as undiluted blood or milk. The presence of cleaning agents, disinfectants, pesticides, or antibiotics can have detrimental impacts on treatment processes.

Activated Sludge Process

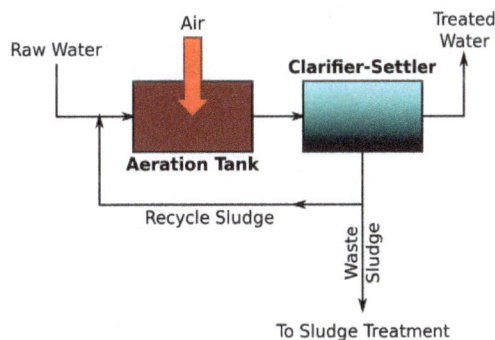

A generalized diagram of an activated sludge process.

Activated sludge is a biochemical process for treating sewage and industrial wastewater that uses air (or oxygen) and microorganisms to biologically oxidize organic pollutants, producing a waste sludge (or floc) containing the oxidized material. In general, an activated sludge process includes:

- An aeration tank where air (or oxygen) is injected and thoroughly mixed into the wastewater.

- A settling tank (usually referred to as a clarifier or "settler") to allow the waste sludge to settle. Part of the waste sludge is recycled to the aeration tank and the remaining waste sludge is removed for further treatment and ultimate disposal.

Trickling Filter Process

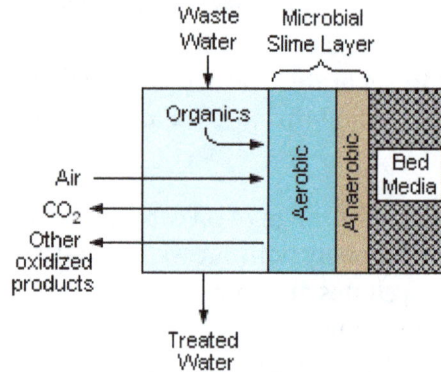

Image 1: A schematic cross-section of the contact face of the bed media in a trickling filter

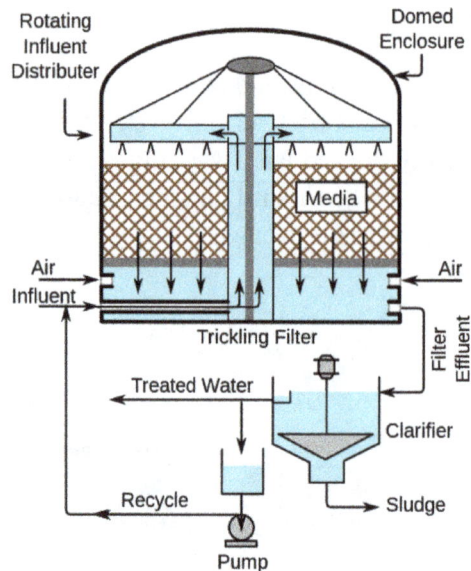

A typical complete trickling filter system

A trickling filter consists of a bed of rocks, gravel, slag, peat moss, or plastic media over which wastewater flows downward and contacts a layer (or film) of microbial slime covering the bed media. Aerobic conditions are maintained by forced air flowing through the bed or by natural convection of air. The process involves adsorption of organic compounds in the wastewater by the microbial slime layer, diffusion of air into the slime layer to provide the oxygen required for the biochemical oxidation of the organic compounds. The end products include carbon dioxide gas, water and other products of the oxidation. As the slime layer thickens, it becomes difficult for the air to penetrate the layer and an inner anaerobic layer is formed.

The fundamental components of a complete trickling filter system are:

- A bed of filter medium upon which a layer of microbial slime is promoted and developed.

- An enclosure or a container which houses the bed of filter medium.

- A system for distributing the flow of wastewater over the filter medium.

- A system for removing and disposing of any sludge from the treated effluent.

The treatment of sewage or other wastewater with trickling filters is among the oldest and most well characterized treatment technologies.

A trickling filter is also often called a *trickle filter*, *trickling biofilter*, *biofilter*, *biological filter* or *biological trickling filter*.

Treatment of Other Organics

Synthetic organic materials including solvents, paints, pharmaceuticals, pesticides, products from coke production and so forth can be very difficult to treat. Treatment methods are often specific to the material being treated. Methods include advanced oxidation processing, distillation, adsorption, vitrification, incineration, chemical immobilisation or landfill disposal. Some materials such as some detergents may be capable of biological degradation and in such cases, a modified form of wastewater treatment can be used.

Treatment of Acids and Alkalis

Acids and alkalis can usually be neutralised under controlled conditions. Neutralisation frequently produces a precipitate that will require treatment as a solid residue that may also be toxic. In some cases, gases may be evolved requiring treatment for the gas stream. Some other forms of treatment are usually required following neutralisation.

Waste streams rich in hardness ions as from de-ionisation processes can readily lose the hardness ions in a buildup of precipitated calcium and magnesium salts. This precipitation process can cause severe *furring* of pipes and can, in extreme cases, cause the blockage of disposal pipes. A 1 metre diameter industrial marine discharge pipe serving a major chemicals complex was blocked by such salts in the 1970s. Treatment is by concentration of de-ionisation waste waters and disposal to landfill or by careful pH management of the released wastewater.

Treatment of Toxic Materials

Toxic materials including many organic materials, metals (such as zinc, silver, cadmium, thallium, etc.) acids, alkalis, non-metallic elements (such as arsenic or selenium) are generally resistant to biological processes unless very dilute. Metals can often be precipitated out by changing the pH or by treatment with other chemicals. Many, however, are resistant to treatment or mitigation and may require concentration followed by landfilling or recycling. Dissolved organics can be *incinerated* within the wastewater by the advanced oxidation process.

Agricultural Wastewater Treatment

Agricultural wastewater treatment is the treatment of wastewaters produced in the course of agricultural activities. Agriculture is a highly intensified industry in many parts of the world, producing a range of wastewaters requiring a variety of treatment technologies and management practices.

Riparian buffer lining a creek in Iowa

Nonpoint Source Pollution

Nonpoint source pollution from farms is caused by surface runoff from fields during rain storms. Agricultural runoff is a major source of pollution, in some cases the only source, in many watersheds.

Sediment Runoff

Highly erodible soils on a farm in Iowa

Soil washed off fields is the largest source of agricultural pollution in the United States. Excess sediment causes high levels of turbidity in water bodies, which can inhibit growth of aquatic plants, clog fish gills and smother animal larvae.

Farmers may utilize erosion controls to reduce runoff flows and retain soil on their fields. Common techniques include:

- contour ploughing

- crop mulching

- crop rotation

- planting perennial crops

- installing riparian buffers.

Nutrient Runoff

Manure spreader

Nitrogen and phosphorus are key pollutants found in runoff, and they are applied to farmland in several ways, such as in the form of commercial fertilizer, animal manure, or municipal or industrial wastewater (effluent) or sludge. These chemicals may also enter runoff from crop residues, irrigation water, wildlife, and atmospheric deposition.

Farmers can develop and implement nutrient management plans to mitigate impacts on water quality by:

- mapping and documenting fields, crop types, soil types, water bodies

- developing realistic crop yield projections

- conducting soil tests and nutrient analyses of manures and/or sludges applied

- identifying other significant nutrient sources (e.g., irrigation water)

- evaluating significant field features such as highly erodible soils, subsurface drains, and shallow aquifers

- applying fertilizers, manures, and/or sludges based on realistic yield goals and using precision agriculture techniques.

Pesticides

Aerial application (crop dusting) of pesticides over a soybean field in the U.S.

Pesticides are widely used by farmers to control plant pests and enhance production, but chemical pesticides can also cause water quality problems. Pesticides may appear in surface water due to:

- direct application (e.g. aerial spraying or broadcasting over water bodies)

- runoff during rain storms

- aerial drift (from adjacent fields).

Some pesticides have also been detected in groundwater.

Farmers may use Integrated Pest Management (IPM) techniques (which can include biological pest control) to maintain control over pests, reduce reliance on chemical pesticides, and protect water quality.

There are few safe ways of disposing of pesticide surpluses other than through containment in well managed landfills or by incineration. In some parts of the world, spraying on land is a permitted method of disposal.

Point Source Pollution

Farms with large livestock and poultry operations, such as factory farms, can be a major source of point source wastewater. In the United States, these facilities are called *concentrated animal feeding operations* or *confined animal feeding operations* and are being subject to increasing government regulation.

Animal Wastes

Confined Animal Feeding Operation in the United States

The constituents of animal wastewater typically contain

- Strong organic content — much stronger than human sewage

- High solids concentration

- High nitrate and phosphorus content

- Antibiotics

- Synthetic hormones

- Often high concentrations of parasites and their eggs

- Spores of *Cryptosporidium* (a protozoan) resistant to drinking water treatment processes

- Spores of *Giardia*

- Human pathogenic bacteria such as *Brucella* and *Salmonella*

Animal wastes from cattle can be produced as solid or semisolid manure or as a liquid slurry. The production of slurry is especially common in housed dairy cattle.

Treatment

Whilst solid manure heaps outdoors can give rise to polluting wastewaters from runoff, this type of waste is usually relatively easy to treat by containment and/or covering of the heap.

Animal slurries require special handling and are usually treated by containment in lagoons before disposal by spray or trickle application to grassland. Constructed wetlands are sometimes used to facilitate treatment of animal wastes, as are anaerobic lagoons. Excessive application or application to sodden land or insufficient land area can result in direct runoff to watercourses, with the potential for causing severe pollution. Application of slurries to land overlying aquifers can result in direct contamination or, more commonly, elevation of nitrogen levels as nitrite or nitrate.

The disposal of any wastewater containing animal waste upstream of a drinking water intake can pose serious health problems to those drinking the water because of the highly resistant spores present in many animals that are capable of causing disabling disease in humans. This risk exists even for very low-level seepage via shallow surface drains or from rainfall run-off.

Some animal slurries are treated by mixing with straws and composted at high temperature to produce a bacteriologically sterile and friable manure for soil improvement.

Piggery Waste

Hog confinement barn or piggery

Piggery waste is comparable to other animal wastes and is processed as for general animal waste, except that many piggery wastes contain elevated levels of copper that can be toxic in the natural environment. The liquid fraction of the waste is frequently separated off and re-used in the piggery to avoid the prohibitively expensive costs of disposing of copper-rich liquid. Ascarid worms and their eggs are also common in piggery waste and can infect humans if wastewater treatment is ineffective.

Silage Liquor

Fresh or wilted grass or other green crops can be made into a semi-fermented product called silage which can be stored and used as winter forage for cattle and sheep. The production of silage often involves the use of an acid conditioner such as sulfuric acid or formic acid. The process of silage making frequently produces a yellow-brown strongly smelling liquid which is very rich in simple sugars, alcohol, short-chain organic acids and silage conditioner. This liquor is one of the most polluting organic substances known. The volume of silage liquor produced is generally in proportion to the moisture content of the ensiled material.

Treatment

Silage liquor is best treated through prevention by wilting crops well before silage making. Any silage liquor that is produced can be used as part of the food for pigs. The most effective treatment is by containment in a slurry lagoon and by subsequent spreading on land following substantial dilution with slurry. Containment of silage liquor on its own can cause structural problems in concrete pits because of the acidic nature of silage liquor.

Milking Parlour (Dairy Farming) Wastes

Although milk has a deserved reputation as an important and valuable food product, its presence in wastewaters is highly polluting because of its organic strength, which can lead to very rapid de-oxygenation of receiving waters. Milking parlour wastes also contain large volumes of wash-down water, some animal waste together with cleaning and disinfection chemicals.

Treatment

Milking parlour wastes are often treated in admixture with human sewage in a local sewage treatment plant. This ensures that disinfectants and cleaning agents are sufficiently diluted and amenable to treatment. Running milking wastewaters into a farm slurry lagoon is a possible option although this tends to consume lagoon capacity very quickly. Land spreading is also a treatment option.

Slaughtering Waste

Wastewater from slaughtering activities is similar to milking parlour waste although considerably stronger in its organic composition and therefore potentially much more polluting.

Treatment

As for milking parlour waste

Vegetable Washing Water

Washing of vegetables produces large volumes of water contaminated by soil and vegetable pieces. Low levels of pesticides used to treat the vegetables may also be present together with moderate levels of disinfectants such as chlorine.

Treatment

Most vegetable washing waters are extensively recycled with the solids removed by settlement and filtration. The recovered soil can be returned to the land.

Firewater

Although few farms plan for fires, fires are nevertheless more common on farms than on many other industrial premises. Stores of pesticides, herbicides, fuel oil for farm machinery and fertilizers can all help promote fire and can all be present in environmentally lethal quantities in firewater from fire fighting at farms.

Treatment

All farm environmental management plans should allow for containment of substantial quantities of firewater and for its subsequent recovery and disposal by specialist disposal companies. The concentration and mixture of contaminants in firewater make them unsuited to any treatment method available on the farm. Even land spreading has produced severe taste and odour problems for downstream water supply companies in the past.

Sewage Treatment

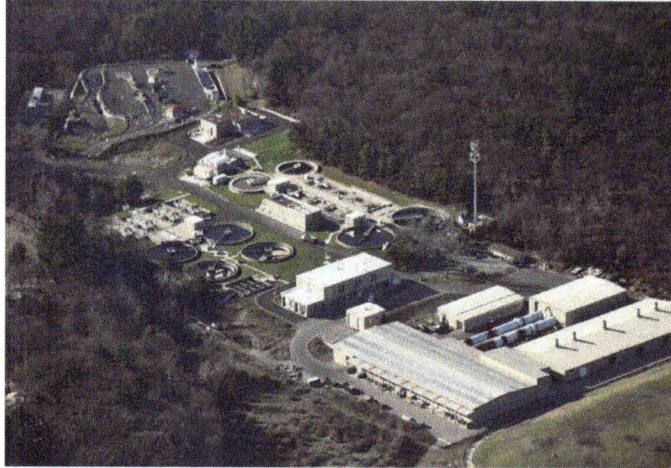

Wastewater treatment plant in Massachusetts, United States

Sewage treatment is the process of removing contaminants from wastewater, primarily from household sewage. It includes physical, chemical, and biological processes to remove these contaminants and produce environmentally safe treated wastewater (or treated effluent). A by-product of sewage treatment is usually a semi-solid waste or slurry, called sewage sludge, that has to undergo further treatment before being suitable for disposal or land application.

Sewage treatment may also be referred to as wastewater treatment, although the latter is a broader term which can also be applied to purely industrial wastewater. For most cities, the sewer system will also carry a proportion of industrial effluent to the sewage treatment plant which has usually received pretreatment at the factories themselves to reduce the pollutant load. If the sewer system is a combined sewer then it will also carry urban runoff (stormwater) to the sewage treatment plant.

Terminology

The term "sewage treatment plant" (or "sewage treatment works" in some countries) is nowadays often replaced with the term "wastewater treatment plant".

Sewage can be treated close to where the sewage is created, which may be called a "decentralized" system or even an "on-site" system (in septic tanks, biofilters or aerobic treatment systems). Alternatively, sewage can be collected and transported by a network of pipes and pump stations to a municipal treatment plant. This is called a "centralized" system although the borders between decentralized and centralized can be variable. For this reason, the terms "semi-decentralized" and "semi-centralized" are also being used.

Origins of Sewage

Sewage is generated by residential, institutional, commercial and industrial establishments. It includes household waste liquid from toilets, baths, showers, kitchens, and sinks draining into sewers. In many areas, sewage also includes liquid waste from industry and commerce. The separation and draining of household waste into greywater and blackwater is becoming more common in the developed world, with treated greywater being permitted to be used for watering plants or recycled for flushing toilets.

Sewage Mixing with Rainwater

Sewage may include stormwater runoff or urban runoff. Sewerage systems capable of handling storm water are known as combined sewer systems. This design was common when urban sewerage systems were first developed, in the late 19th and early 20th centuries. Combined sewers require much larger and more expensive treatment facilities than sanitary sewers. Heavy volumes of storm runoff may overwhelm the sewage treatment system, causing a spill or overflow. Sanitary sewers are typically much smaller than combined sewers, and they are not designed to transport stormwater. Backups of raw sewage can occur if excessive infiltration/inflow (dilution by stormwater and/or groundwater) is allowed into a sanitary sewer system. Communities that have urbanized in the mid-20th century or later generally have built separate systems for sewage (sanitary sewers) and stormwater, because precipitation causes widely varying flows, reducing sewage treatment plant efficiency.

As rainfall travels over roofs and the ground, it may pick up various contaminants including soil particles and other sediment, heavy metals, organic compounds, animal waste, and oil and grease. Some jurisdictions require stormwater to receive some level of treatment before being discharged directly into waterways. Examples of treatment processes used for stormwater include retention basins, wetlands, buried vaults with various kinds of media filters, and vortex separators (to remove coarse solids).

Industrial Effluent

In highly regulated developed countries, industrial effluent usually receives at least pretreatment if not full treatment at the factories themselves to reduce the pollutant load, before discharge to the sewer. This process is called industrial wastewater treatment. The same does not apply to many developing countries where industrial effluent is more likely to enter the sewer if it exists, or even the receiving water body, without pretreatment.

Industrial wastewater may contain pollutants which cannot be removed by conventional sewage treatment. Also, variable flow of industrial waste associated with production cycles may upset the population dynamics of biological treatment units, such as the activated sludge process.

Process Steps

Overview

Sewage collection and treatment is typically subject to local, state and federal regulations and standards.

Treating wastewater has the aim to produce an effluent that will do as little harm as possible when discharged to the surrounding environment, thereby preventing pollution compared to releasing untreated wastewater into the environment.

Sewage treatment generally involves three stages, called primary, secondary and tertiary treatment.

- *Primary treatment* consists of temporarily holding the sewage in a quiescent basin where heavy solids can settle to the bottom while oil, grease and lighter solids float to the surface. The settled and floating materials are removed and the remaining liquid may be discharged or subjected to secondary treatment. Some sewage treatment plants that are connected to a combined sewer system have a bypass arrangement after the primary treatment unit. This means that during very heavy rainfall events, the secondary and tertiary treatment systems can be bypassed to protect them from hydraulic overloading, and the mixture of sewage and stormwater only receives primary treatment.

- *Secondary treatment* removes dissolved and suspended biological matter. Secondary treatment is typically performed by indigenous, water-borne micro-organisms in a managed habitat. Secondary treatment may require a separation process to remove the micro-organisms from the treated water prior to discharge or tertiary treatment.

- *Tertiary treatment* is sometimes defined as anything more than primary and secondary treatment in order to allow rejection into a highly sensitive or fragile ecosystem (estuaries, low-flow rivers, coral reefs,...). Treated water is sometimes disinfected chemically or physically (for example, by lagoons and microfiltration) prior to discharge into a stream, river, bay, lagoon or wetland, or it can be used for the irrigation of a golf course, green way or park. If it is sufficiently clean, it can also be used for groundwater recharge or agricultural purposes.

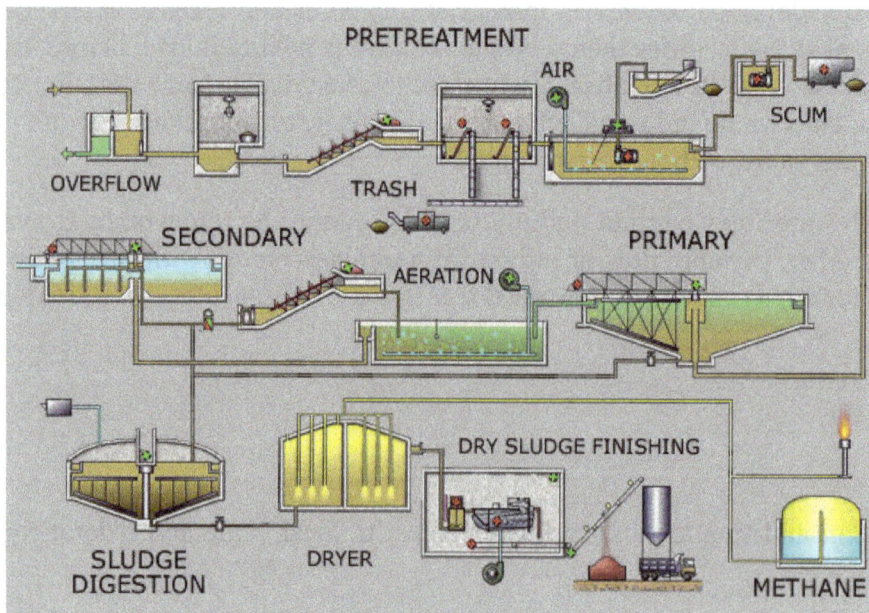

Simplified process flow diagram for a typical large-scale treatment plant

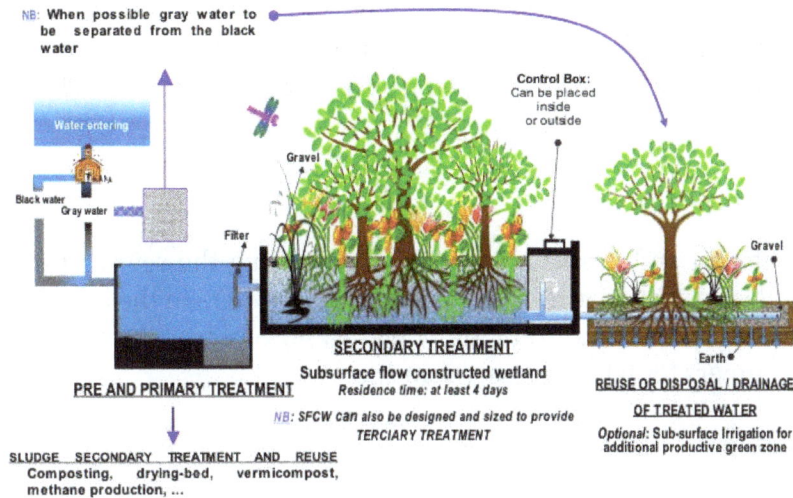

Process flow diagram for a typical treatment plant via subsurface flow constructed wetlands (SFCW)

Pretreatment

Pretreatment removes all materials that can be easily collected from the raw sewage before they damage or clog the pumps and sewage lines of primary treatment clarifiers. Objects commonly removed during pretreatment include trash, tree limbs, leaves, branches, and other large objects.

The influent in sewage water passes through a bar screen to remove all large objects like cans, rags, sticks, plastic packets etc. carried in the sewage stream. This is most commonly done with an automated mechanically raked bar screen in modern plants serving large populations, while in smaller or less modern plants, a manually cleaned screen may be used. The raking action of a mechanical bar screen is typically paced according to the accumulation on the bar screens and/or flow rate. The solids are collected and later disposed in a landfill, or incinerated. Bar screens or mesh screens of varying sizes may be used to optimize solids removal. If gross solids are not removed, they become entrained in pipes and moving parts of the treatment plant, and can cause substantial damage and inefficiency in the process.[9]

Grit Removal

Pretreatment may include a sand or grit channel or chamber, where the velocity of the incoming sewage is adjusted to allow the settlement of sand, grit, stones, and broken glass. These particles are removed because they may damage pumps and other equipment. For small sanitary sewer systems, the grit chambers may not be necessary, but grit removal is desirable at larger plants. Grit chambers come in 3 types: horizontal grit chambers, aerated grit chambers and vortex grit chambers. The process is called sedimentation.

Flow Equalization

Clarifiers and mechanized secondary treatment are more efficient under uniform flow conditions. Equalization basins may be used for temporary storage of diurnal or wet-weather flow peaks. Basins provide a place to temporarily hold incoming sewage during plant maintenance and a means of diluting and distributing batch discharges of toxic or high-strength waste which might otherwise inhibit biological secondary treatment (including portable toilet waste, vehicle holding tanks, and septic tank pumpers). Flow equalization basins require variable discharge control, typically include provisions for bypass and cleaning, and may also include aerators. Cleaning may be easier if the basin is downstream of screening and grit removal.

Fat and Grease Removal

In some larger plants, fat and grease are removed by passing the sewage through a small tank where skimmers collect the fat floating on the surface. Air blowers in the base of the tank may also be used to help recover the fat as a froth. Many plants, however, use primary clarifiers with mechanical surface skimmers for fat and grease removal.

Primary Treatment

Primary treatment tanks in Oregon, USA.

In the primary sedimentation stage, sewage flows through large tanks, commonly called "pre-settling basins", "primary sedimentation tanks" or "primary clarifiers". The tanks are used to settle sludge while grease and oils rise to the surface and are skimmed off. Primary settling tanks are usually equipped with mechanically driven scrapers that continually drive the collected sludge towards a hopper in the base of the tank where it is pumped to sludge treatment facilities.[9-11] Grease and oil from the floating material can sometimes be recovered for saponification (soap making).

Secondary Treatment

Secondary treatment is designed to substantially degrade the biological content of the sewage which are derived from human waste, food waste, soaps and detergent. The majority of municipal plants treat the settled sewage liquor using aerobic biological processes. To be effective, the biota require both oxygen and food to live. The bacteria and protozoa consume biodegradable soluble organic contaminants (e.g. sugars, fats, organic short-chain carbon molecules, etc.) and bind much of the less soluble fractions into floc. Secondary treatment systems are classified as *fixed-film* or *suspended-growth* systems.

- Fixed-film or attached growth systems include trickling filters, bio-towers, and rotating biological contactors, where the biomass grows on media and the sewage passes over its surface.[11-13] The fixed-film principle has further developed into Moving Bed Biofilm Reactors (MBBR) and Integrated Fixed-Film Activated Sludge (IFAS) processes. An MBBR system typically requires a smaller footprint than suspended-growth systems.

- Suspended-growth systems include activated sludge, where the biomass is mixed with the sewage and can be operated in a smaller space than trickling filters that treat the same amount of water. However, fixed-film systems are more able to cope with drastic changes in the amount of biological material and can provide higher removal rates for organic material and suspended solids than suspended growth systems.

Secondary Sedimentation

Secondary clarifier at a rural treatment plant.

Some secondary treatment methods include a secondary clarifier to settle out and separate biological floc or filter material grown in the secondary treatment bioreactor.

List of Process Types

- Activated sludge
- Aerated lagoon
- Aerobic granulation
- Constructed wetland
- Membrane bioreactor
- Rotating biological contactor
- Sequencing batch reactor
- Trickling filter

To use less space, treat difficult waste, and intermittent flows, a number of designs of hybrid treatment plants have been produced. Such plants often combine at least two stages of the three main treatment stages into one combined stage. In the UK, where a large number of wastewater treatment plants serve small populations, package plants are a viable alternative to building a large structure for each process stage. In the US, package plants are typically used in rural areas, highway rest stops and trailer parks.

Tertiary Treatment

The purpose of tertiary treatment is to provide a final treatment stage to further improve the effluent quality before it is discharged to the receiving environment (sea, river, lake, wet lands, ground, etc.). More than one tertiary treatment process may be used at any treatment plant. If disinfection is practised, it is always the final process. It is also called "effluent polishing."

Filtration

Sand filtration removes much of the residual suspended matter.[22–23] Filtration over activated carbon, also called *carbon adsorption,* removes residual toxins.

Lagoons or Ponds

A sewage treatment plant and lagoon in Everett, Washington, United States.

Lagoons or ponds provide settlement and further biological improvement through storage in large man-made ponds or lagoons. These lagoons are highly aerobic and colonization by native macrophytes, especially reeds, is often encouraged. Small filter feeding invertebrates such as *Daphnia* and species of *Rotifera* greatly assist in treatment by removing fine particulates.

Biological Nutrient Removal

Biological nutrient removal (BNR) is regarded by some as a type of secondary treatment process, and by others as a tertiary (or "advanced") treatment process.

Wastewater may contain high levels of the nutrients nitrogen and phosphorus. Excessive release to the environment can lead to a buildup of nutrients, called eutrophication, which can in turn encourage the overgrowth of weeds, algae, and cyanobacteria (blue-green algae). This may cause an algal bloom, a rapid growth in the population of algae. The algae numbers are unsustainable and eventually most of them die. The decomposition of the algae by bacteria uses up so much of the oxygen in the water that most or all of the animals die, which creates more organic matter for the bacteria to decompose. In addition to causing deoxygenation, some algal species produce toxins that contaminate drinking water supplies. Different treatment processes are required to remove nitrogen and phosphorus.

Nitrogen Removal

Nitrogen is removed through the biological oxidation of nitrogen from ammonia to nitrate (nitrification), followed by denitrification, the reduction of nitrate to nitrogen gas. Nitrogen gas is released to the atmosphere and thus removed from the water.

Nitrification itself is a two-step aerobic process, each step facilitated by a different type of bacteria. The oxidation of ammonia (NH_3) to nitrite (NO_2^-) is most often facilitated by *Nitrosomonas* spp. ("nitroso" referring to the formation of a nitroso functional group). Nitrite oxidation to nitrate (NO_3^-), though traditionally believed to be facilitated by *Nitrobacter* spp. (nitro referring the formation of a nitro functional group), is now known to be facilitated in the environment almost exclusively by *Nitrospira* spp.

Denitrification requires anoxic conditions to encourage the appropriate biological communities to form. It is facilitated by a wide diversity of bacteria. Sand filters, lagooning and reed beds can all be used to reduce nitrogen, but the activated sludge process (if designed well) can do the job the most easily.[17-18] Since denitrification is the reduction of nitrate to dinitrogen (molecular nitrogen) gas, an electron donor is needed. This can be, depending on the waste water, organic matter (from feces), sulfide, or an added donor like methanol. The sludge in the anoxic tanks (denitrification tanks) must be mixed well (mixture of recirculated mixed liquor, return activated sludge [RAS], and raw influent) e.g. by using submersible mixers in order to achieve the desired denitrification.

Sometimes the conversion of toxic ammonia to nitrate alone is referred to as tertiary treatment.

Over time, different treatment configurations have evolved as denitrification has become more sophisticated. An initial scheme, the Ludzack-Ettinger Process, placed an anoxic treatment zone before the aeration tank and clarifier, using the return activated sludge (RAS) from the clarifier as a nitrate source. Influent wastewater (either raw or as effluent from primary clarification) serves as the electron source for the facultative bacteria to metabolize carbon, using the inorganic nitrate as a source of oxygen instead of dissolved molecular oxygen. This denitrification scheme was naturally limited to the amount of soluble nitrate present in the RAS. Nitrate reduction was limited because RAS rate is limited by the performance of the clarifier.

The "Modified Ludzak-Ettinger Process" (MLE) is an improvement on the original concept, for it recycles mixed liquor from the discharge end of the aeration tank to the head of the anoxic tank to provide a consistent source of soluble nitrate for the facultative bacteria. In this instance, raw wastewater continues to provide the electron source, and sub-surface mixing maintains the bacteria in contact with both electron source and soluble nitrate in the absence of dissolved oxygen.

Many sewage treatment plants use centrifugal pumps to transfer the nitrified mixed liquor from the aeration zone to the anoxic zone for denitrification. These pumps are often referred to as *Internal Mixed Liquor Recycle* (IMLR) pumps. IMLR may be 200% to 400% the flow rate of influent wastewater (Q.) This is in addition to Return Activated Sludge (RAS) from secondary clarifiers, which may be 100% of Q. (Therefore, the hydraulic capacity of the tanks in such a system should handle at least 400% of annual average design flow (AADF.) At times, the raw or primary effluent wastewater must be carbon-supplemented by the addition of methanol, acetate, or simple food waste (molasses, whey, plant starch) to improve the treatment efficiency. These carbon additions should be accounted for in the design of a treatment facility's organic loading.

Further modifications to the MLE were to come: Bardenpho and Biodenipho processes include additional anoxic and oxidative processes to further polish the conversion of nitrate ion to molecular nitrogen gas. Use of an anaerobic tank following the initial anoxic process allows for luxury uptake of phosphorus by bacteria, thereby biologically reducing orthophosphate ion in the treated wastewater. Even newer improvements, such as Anammox Process, interrupt the formation of nitrate at the nitrite stage of nitrification, shunting nitrite-rich mixed liquor activated sludge to treatment where nitrite is then converted to molecular nitrogen gas, saving energy, alkalinity, and secondary carbon sourcing. Anammox™ (ANaerobic AMMonia OXidation) works by artificially extending detention time and preserving denitrifiying bacteria through the use of substrate added to the mixed liquor and continuously recycled from it prior to secondary clarification. Many other proprietary schemes are being deployed, including DEMON™, Sharon-ANAMMOX™, ANITA-Mox™, and DeAmmon™. The bacteria Brocadia anammoxidans can remove ammonium from waste water through anaerobic oxidation of ammonium to hydrazine, a form of rocket fuel.

Phosphorus Removal

Every adult human excretes between 200 and 1000 grams of phosphorus annually. Studies of United States sewage in the late 1960s estimated mean per capita contributions of 500 grams in urine and feces, 1000 grams in synthetic detergents, and lesser variable amounts used as corrosion and scale control chemicals in water supplies. Source control via alternative detergent formulations has subsequently reduced the largest contribution, but the content of urine and feces will remain unchanged. Phosphorus removal is important as it is a limiting nutrient for algae growth in many fresh water systems. It is also particularly important for water reuse systems where high phosphorus concentrations may lead to fouling of downstream equipment such as reverse osmosis.

Phosphorus can be removed biologically in a process called enhanced biological phosphorus removal. In this process, specific bacteria, called polyphosphate-accumulating organisms (PAOs), are selectively enriched and accumulate large quantities of phosphorus within their cells (up to 20 percent of their mass). When the biomass enriched in these bacteria is separated from the treated water, these biosolids have a high fertilizer value.

Phosphorus removal can also be achieved by chemical precipitation, usually with salts of iron (e.g. ferric chloride), aluminum (e.g. alum), or lime. This may lead to excessive sludge production as hydroxides precipitates and the added chemicals can be expensive. Chemical phosphorus removal requires significantly smaller equipment footprint than biological removal, is easier to operate and is often more reliable than biological phosphorus removal. Another method for phosphorus removal is to use granular laterite.

Once removed, phosphorus, in the form of a phosphate-rich sewage sludge, may be dumped in a landfill or used as fertilizer. In the latter case, the treated sewage sludge is also sometimes referred to as biosolids.

Disinfection

The purpose of disinfection in the treatment of waste water is to substantially reduce the number of microorganisms in the water to be discharged back into the environment for the later use of drinking, bathing, irrigation, etc. The effectiveness of disinfection depends on the quality of the water being treated (e.g., cloudiness, pH, etc.), the type of disinfection being used, the disinfectant dosage (concentration and time), and other environmental variables. Cloudy water will be treated less successfully, since solid matter can shield organisms, especially from ultraviolet light or if contact times are low. Generally, short contact times, low doses and high flows all militate against effective disinfection. Common methods of disinfection include ozone, chlorine, ultraviolet light, or sodium hypochlorite.[16] Chloramine, which is used for drinking water, is not used in the treatment of waste water because of its persistence. After multiple steps of disinfection, the treated water is ready to be released back into the water cycle by means of the nearest body of water or agriculture. Afterwards, the water can be transferred to reserves for everyday human uses.

Chlorination remains the most common form of waste water disinfection in North America due to its low cost and long-term history of effectiveness. One disadvantage is that chlorination of residual organic material can generate chlorinated-organic compounds that may be carcinogenic or harmful to the environment. Residual chlorine or chloramines may also be capable of chlorinating organic material in the natural aquatic environment. Further, because residual chlorine is toxic to aquatic species, the treated effluent must also be chemically dechlorinated, adding to the complexity and cost of treatment.

Ultraviolet (UV) light can be used instead of chlorine, iodine, or other chemicals. Because no chemicals are used, the treated water has no adverse effect on organisms that later consume it, as may be the case with other methods. UV radiation causes damage to the genetic structure of bacteria, viruses, and other pathogens, making them incapable of reproduction. The key disadvantages of UV disinfection are the need for frequent lamp maintenance and replacement and the need for a highly treated effluent to ensure that the target microorganisms are not shielded from the UV radiation (i.e., any solids present in the treated effluent may protect microorganisms from the UV light). In the United Kingdom, UV light is becoming the most common means of disinfection because of the concerns about the impacts of chlorine in chlorinating residual organics in the wastewater and in chlorinating organics in the receiving water. Some sewage treatment systems in Canada and the US also use UV light for their effluent water disinfection.

Ozone (O_3) is generated by passing oxygen (O_2) through a high voltage potential resulting in a third oxygen atom becoming attached and forming O_3. Ozone is very unstable and reactive and oxidizes most organic material it comes in contact with, thereby destroying many pathogenic microorganisms. Ozone is considered to be safer than chlorine because, unlike chlorine which has to be stored on site (highly poisonous in the event of an accidental release), ozone is generated on-site as needed. Ozonation also produces fewer disinfection by-products than chlorination. A disadvantage of ozone disinfection is the high cost of the ozone generation equipment and the requirements for special operators.

Fourth Treatment Stage

Micropollutants such as pharmaceuticals, ingredients of household chemicals, chemicals used in small businesses or industries, environmental persistent pharmaceutical pollutant (EPPP) or pesticides may not be eliminated in the conventional treatment process (primary, secondary and tertiary treatment) and therefore lead to water pollution. Although concentrations of those substances and their decompostion products are quite low, there is still a chance to harm aquatic organisms. For pharmaceuticals, the following substances have been identified as "toxicologically relevant": substances with endocrine disrupting effects, genotoxic substances and substances that enhance the development of bacterial resistances. They mainly belong to the group of environmental persistent pharmaceutical pollutants. Techniques for elimination of micropollutants via a fourth treatment stage during sewage treatment are being tested in Germany, Switzerland and the Netherlands. However, since those techniques are still costly, they are not yet applied on a regular basis. Such process steps mainly consist of activated carbon filters that adsorb the micropollutants. Ozone can also be applied as an oxidative method. Also the use of enzymes such as the enzyme laccase is under investigation. A new concept which could provide an energy-efficient treatment of micropollutants could be the use of laccase secreting fungi cultivated at a wastewater treatment plant to degrade micropollutants and at the same time to provide enzymes at a cathode of a microbial biofuel cells. Microbial biofuel cells are investigated for their property to treat organic matter in wastewater.

To reduce pharmaceuticals in water bodies, also "source control" measures are under investigation, such as innovations in drug development or more responsible handling of drugs.

Odor Control

Odors emitted by sewage treatment are typically an indication of an anaerobic or "septic" condition. Early stages of processing will tend to produce foul-smelling gases, with hydrogen sulfide being most common in generating complaints. Large process plants in urban areas will often treat the odors with carbon reactors, a contact media with bio-slimes, small doses of chlorine, or circulating fluids to biologically capture and metabolize the noxious gases. Other methods of odor control exist, including addition of iron salts, hydrogen peroxide, calcium nitrate, etc. to manage hydrogen sulfide levels.

High-density solids pumps are suitable for reducing odors by conveying sludge through hermetic closed pipework.

Energy Requirements

For conventional sewage treatment plants, around 30 percent of the annual operating costs is usually required for energy. The energy requirements vary with type of treatment process as well as wastewater load. For example, constructed wetlands have a lower energy requirement than activated sludge plants, as less energy is required for the aeration step. Sewage treatment plants that produce biogas in their sewage sludge treatment process with anaerobic digestion can produce enough energy to meet most of the energy needs of the sewage treatment plant itself.

In conventional secondary treatment processes, most of the electricity is used for aeration, pumping systems and equipment for the dewatering and drying of sewage sludge. Advanced wastewater treatment plants, e.g. for nutrient removal, require more energy than plants that only achieve primary or secondary treatment.

Sludge Treatment and Disposal

The sludges accumulated in a wastewater treatment process must be treated and disposed of in a safe and effective manner. The purpose of digestion is to reduce the amount of organic matter and the number of disease-causing microorganisms present in the solids. The most common treatment options include anaerobic digestion, aerobic digestion, and composting. Incineration is also used, albeit to a much lesser degree.

Sludge treatment depends on the amount of solids generated and other site-specific conditions. Composting is most often applied to small-scale plants with aerobic digestion for mid-sized operations, and anaerobic digestion for the larger-scale operations.

The sludge is sometimes passed through a so-called pre-thickener which de-waters the sludge. Types of pre-thickeners include centrifugal sludge thickeners rotary drum sludge thickeners and belt filter presses. Dewatered sludge may be incinerated or transported offsite for disposal in a landfill or use as an agricultural soil amendment.

Environment Aspects

The outlet of the Karlsruhe sewage treatment plant flows into the Alb.

Many processes in a wastewater treatment plant are designed to mimic the natural treatment processes that occur in the environment, whether that environment is a natural water body or the ground. If not overloaded, bacteria in the environment will consume organic contaminants, although this will reduce the levels of oxygen in the water and may significantly change the overall ecology of the receiving water. Native bacterial populations feed on the organic contaminants, and the numbers of disease-causing microorganisms are reduced by natural environmental conditions such as predation or exposure to ultraviolet radiation. Consequently, in cases where the receiving environment provides a high level of dilution, a high degree of wastewater treatment may not be required. However, recent evidence has demonstrated that very low levels of specific contaminants in wastewater, including hormones (from animal husbandry and residue from human hormonal contraception methods) and synthetic materials such as phthalates that mimic hormones in their action, can have an unpredictable adverse impact on the natural biota and potentially on humans if the water is re-used for drinking water. In the US and EU, uncontrolled discharges of wastewater to the environment are not permitted under law, and strict water quality requirements are to be met, as clean drinking water is essential. A sig-nificant threat in the coming decades will be the increasing uncontrolled discharges of wastewater within rapidly developing countries.

Effects on Biology

Sewage treatment plants can have multiple effects on nutrient levels in the water that the treated sewage flows into. These nutrients can have large effects on the biological life in the water in contact with the effluent. Stabilization ponds (or sewage treatment ponds) can include any of the following:

- Oxidation ponds, which are aerobic bodies of water usually 1–2 meters in depth that receive effluent from sedimentation tanks or other forms of primary treatment.

    - Dominated by algae

- Polishing ponds are similar to oxidation ponds but receive effluent from an oxidation pond or from a plant with an extended mechanical treatment.

    - Dominated by zooplankton

- Facultative lagoons, raw sewage lagoons, or sewage lagoons are ponds where sewage is added with no primary treatment other than coarse screening. These ponds provide effective treatment when the surface remains aerobic; although anaerobic conditions may develop near the layer of settled sludge on the bottom of the pond.

- Anaerobic lagoons are heavily loaded ponds.

    - Dominated by bacteria

- Sludge lagoons are aerobic ponds, usually 2 to 5 meters in depth, that receive anaerobically digested primary sludge, or activated secondary sludge under water.

    - Upper layers are dominated by algae

Phosphorus limitation is a possible result from sewage treatment and results in flagellate-dominated plankton, particularly in summer and fall.

A phytoplankton study found high nutrient concentrations linked to sewage effluents. High nutrient concentration leads to high chlorophyll a concentrations, which is a proxy for primary production in marine environments. High primary production means high phytoplankton populations and most likely high zooplankton populations, because zooplankton feed on phytoplankton. However, effluent released into marine systems also leads to greater population instability.

The planktonic trends of high populations close to input of treated sewage is contrasted by the bacterial trend. In a study of *Aeromonas* spp. in increasing distance from a wastewater source, greater change in seasonal cycles was found the furthest from the effluent. This trend is so strong that the furthest location studied actually had an inversion of the *Aeromonas* spp. cycle in comparison to that of fecal coliforms. Since there is a main pattern in the cycles that occurred simultaneously at all stations it indicates seasonal factors (temperature, solar radiation, phytoplankton) control of the bacterial population. The effluent dominant species changes from *Aeromonas caviae* in winter to *Aeromonas sobria* in the spring and fall while the inflow dominant species is *Aeromonas caviae*, which is constant throughout the seasons.

Treated Sewage Reuse

With suitable technology, it is possible to reuse sewage effluent for drinking water, although this is usually only done in places with limited water supplies, such as Windhoek and Singapore.

In Israel, about 50 percent of agricultural water use (total use was 1 billion cubic metres in 2008) is provided through reclaimed sewer water. Future plans call for increased use of treated sewer water as well as more desalination plants.

Sewage Treatment in Developing Countries

Few reliable figures exist on the share of the wastewater collected in sewers that is being treated in the world. A global estimate by UNDP and UN-Habitat is that 90% of all wastewater generated is released into the environment untreated. In many developing countries the bulk of domestic and industrial wastewater is discharged without any treatment or after primary treatment only.

In Latin America about 15 percent of collected wastewater passes through treatment plants (with varying levels of actual treatment). In Venezuela, a below average country in South America with respect to wastewater treatment, 97 percent of the country's sewage is discharged raw into the environment. In Iran, a relatively developed Middle Eastern country, the majority of Tehran's population has totally untreated sewage injected to the city's groundwater. However, the construction of major parts of the sewage system, collection and treatment, in Tehran is almost complete, and under development, due to be fully completed by the end of 2012. In Isfahan, Iran's third largest city, sewage treatment was started more than 100 years ago.

Only few cities in sub-Saharan Africa have sewer-based sanitation systems, let alone wastewater treatment plants, an exception being South Africa and – until the late 1990s- Zimbabwe. Instead, most urban residents in sub-Saharan Africa rely on on-site sanitation systems without sewers, such as septic tanks and pit latrines, and faecal sludge management in these cities is an enormous challenge.

History

Basic sewer systems were used for waste removal in ancient Mesopotamia, where vertical shafts carried the waste away into cesspools. Similar systems existed in the Indus Valley civilization in modern-day India and in Ancient Crete and Greece. In the Middle Ages the sewer systems built by the Romans fell into disuse and waste was collected into cesspools that were periodically emptied by workers known as 'rakers' who would often sell it as fertilizer to farmers outside the city.

FARADAY GIVING HIS CARD TO FATHER THAMES;
And we hope the Dirty Fellow will consult the learned Professor.

The Great Stink of 1858 stimulated research into the problem of sewage treatment. In this caricature in *The Times*, Michael Faraday reports to *Father Thames* on the state of the river.

Modern sewage systems were first built in the mid-nineteenth century as a reaction to the exacerbation of sanitary conditions brought on by heavy industrialization and urbanization. Due to the contaminated water supply, cholera outbreaks occurred in 1832, 1849 and 1855 in London, killing tens of thousands of people. This, combined with the Great Stink of 1858, when the smell of untreated human waste in the River Thames became overpowering, and the report into sanitation reform of the Royal Commissioner Edwin Chadwick, led to the Metropolitan Commission of Sewers appointing Sir Joseph Bazalgette to construct a vast underground sewage system for the safe removal of waste. Contrary to Chadwick's recommendations, Bazalgette's system, and others later built in Continental Europe, did not pump the sewage onto farm land for use as fertilizer; it was simply piped to a natural waterway away from population centres, and pumped back into the environment.

Early Attempts

One of the first attempts at diverting sewage for use as a fertilizer in the farm was made by the cotton mill owner James Smith in the 1840s. He experimented with a piped distribution system initially proposed by James Vetch that collected sewage from his factory and pumped it into the outlying farms, and his success was enthusiastically followed by Edwin Chadwick and supported by organic chemist Justus von Liebig.

The idea was officially adopted by the Health of Towns Commission, and various schemes (known as sewage farms) were trialled by different municipalities over the next 50 years. At first, the heavier solids were channeled into ditches on the side of the farm and were covered over when full, but soon flat-bottomed tanks were employed as reservoirs for the sewage; the earliest patent was taken out by William Higgs in 1846 for "tanks or reservoirs in which the contents of sewers and drains from cities, towns and villages are to be collected and the solid animal or vegetable matters therein contained, solidified and dried..." Improvements to the design of the tanks included the introduction of the horizontal-flow tank in the 1850s and the radial-flow tank in 1905. These tanks had to be manually de-sludged periodically, until the introduction of automatic mechanical de-sludgers in the early 1900s.

The precursor to the modern septic tank was the cesspool in which the water was sealed off to prevent contamination and the solid waste was slowly liquified due to anaerobic action; it was invented by L.H Mouras in France in the 1860s. Donald Cameron, as City Surveyor for Exeter patented an improved version in 1895, which he called a 'septic tank'; septic having the meaning of 'bacterial'. These are still in worldwide use, especially in rural areas unconnected to large-scale sewage systems.

Chemical Treatment

Sir Edward Frankland, a distinguished chemist, who demonstrated the possibility of chemically treating sewage in the 1870s.

It was not until the late 19th century that it became possible to treat the sewage by chemically breaking it down through the use of microorganisms and removing the pollutants. Land treatment was also steadily becoming less feasible, as cities grew and the volume of sewage produced could no longer be absorbed by the farmland on the outskirts.

Sir Edward Frankland conducted experiments at the Sewage Farm in Croydon, England, during the 1870s and was able to demonstrate that filtration of sewage through porous gravel produced a nitrified effluent (the ammonia was converted into nitrate) and that the filter remained unclogged over long periods of time. This established the then revolutionary possibility of biological treatment of sewage using a contact bed to oxidize the waste. This concept was taken up by the chief chemist for the London Metropolitan Board of Works, William Libdin, in 1887:

...in all probability the true way of purifying sewage...will be first to separate the sludge, and then turn into neutral effluent... retain it for a sufficient period, during which time it should be fully aerated, and finally discharge it into the stream in a purified condition. This is indeed what is aimed at and imperfectly accomplished on a sewage farm.

From 1885 to 1891 filters working on this principle were constructed throughout the UK and the idea was also taken up in the US at the Lawrence Experiment Station in Massachusetts, where Frankland's work was confirmed. In 1890 the LES developed a 'trickling filter' that gave a much more reliable performance.

Contact beds were developed in Salford, Lancashire and by scientists working for the London City Council in the early 1890s. According to Christopher Hamlin, this was part of a conceptual revolution that replaced the philosophy that saw "sewage purification as the prevention of decomposition with one that tried to facilitate the biological process that destroy sewage naturally."

Contact beds were tanks containing the inert substance, such as stones or slate, that maximized the surface area available for the microbial growth to break down the sewage. The sewage was held in the tank until it was fully decomposed and it was then filtered out into the ground. This method quickly became widespread, especially in the UK, where it was used in Leicester, Sheffield, Manchester and Leeds. The bacterial bed was simultaneously developed by Joseph Corbett as Borough Engineer in Salford and experiments in 1905 showed that his method was superior in that greater volumes of sewage could be purified better for longer periods of time than could be achieved by the contact bed.

The Royal Commission on Sewage Disposal published its eighth report in 1912 that set what became the international standard for sewage discharge into rivers; the '20:30 standard', which allowed 20 mg Biochemical oxygen demand and 30 mg suspended solid per litre.

Rainwater Harvesting

Rainwater capture and storage system at the Monterrey Institute of Technology
and Higher Education, Mexico City.

A cistern for rainwater storage

Rainwater harvesting is the accumulation and deposition of rainwater for reuse on-site, rather than allowing it to run off. Rainwater can be collected from rivers or roofs, and in many places the water collected is redirected to a deep pit (well, shaft, or borehole), a reservoir with percolation, or collected from dew or fog with nets or other tools. Its uses include water for gardens, livestock, irrigation, domestic use with proper treatment, and indoor heating for houses etc. The harvested water can also be used as drinking water, longer-term storage and for other purposes such as groundwater recharge.

Advantages

Rainwater harvesting provides an independent water supply during regional water restrictions and in developed countries is often used to supplement the main supply. It provides water when there is a drought, can help mitigate flooding of low-lying areas, and reduces demand on wells which may enable groundwater levels to be sustained. It also helps in the availability of potable water as rainwater is substantially free of salinity and other salts. Application of rainwater harvesting in urban water system provides a substantial benefit for both water supply and wastewater subsystems by reducing the need for clean water in water distribution system, less generated stormwater in sewer system, as well as a reduction in stormwater runoff polluting freshwater bodies.

There has been a large body of work focused on the development of Life Cycle Assessment and Life Cycle Costing methodologies to assess the level of environmental impacts and money that can be saved by implementing rainwater harvesting systems.

More development and knowledge is required to understand the benefits rainwater harvesting can provide to agriculture. Many countries especially those with an arid environment use rainwater harvesting as a cheap and reliable source of clean water. To enhance irrigation in arid environments, ridges of soil are constructed in order to trap and prevent rainwater from running down hills and slopes. Even in periods of low rainfall, enough water is collected in order for crops to grow. Water can be collected from roofs, dams, and ponds can be constructed in order to hold large quantities of rainwater so that even on days where there is little to no rainfall, there is enough available to irrigate crops.

Quality

The concentration of contaminants is reduced significantly by diverting the initial flow of run-off water to waste. Improved water quality can also be obtained by using a floating draw-off mechanism (rather than from the base of the tank) and by using a series of tanks, with draw from the last in series. Pre-filtration is a common practice used in the industry to ensure that the water entering the tank is free of large sediment. Pre-filtration is important to keep the system healthy.

Conceptually, a water supply system should match the quality of water with the end use. However, in most of the developed world high quality potable water is used for all end uses. This approach wastes money and energy and imposes unnecessary impacts to the environment. Supplying rainwater that has gone through preliminary filtration measures for non-potable water uses, such as toilet flushing, irrigation, and laundry, may be a significant part of a sustainable water management strategy.

System Setup

Rainwater harvesting systems can range in complexity, from systems that can be installed with minimal skills, to automated systems that require advanced setup and installation. The basic Rainwater harvesting system is more of a plumbing job than a technical job as all the outlets from the building terrace are connected through a pipe to an underground tank that stores water.

Systems are ideally sized to meet the water demand throughout the dry season since it must be big enough to support daily water consumption. Specifically, the rainfall capturing area such as a building roof must be large enough to maintain adequate flow. The water storage tank size should be large enough to contain the captured water.

For low-tech systems, there are many low-tech methods used to capture rainwater: rooftop systems, surface water capture, and pumping the rainwater that has already soaked into the ground or captured in reservoirs and storing it into tanks (cisterns).

Before a rainwater harvesting system is built, it is helpful to use digital tools. For instance, if you want to detect if a region has a high rainwater harvesting potential, rainwater harvesting GIS maps can be made using an online interactive tool. Or if you need to estimate how much water is needed to fulfill a community's water needs, the Rain is Gain tool helps with this. Tools like these can save time and money before a commitment to build a system is undertaken, in addition to making the project sustainable and last a long time.

Life Cycle Assessment: Design for Environment

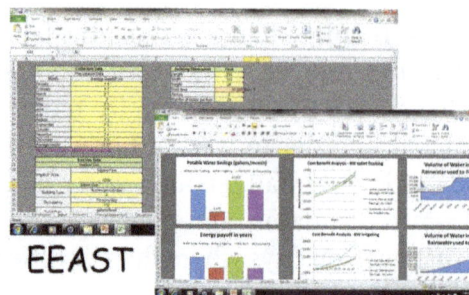

EEAST Model for LCAs of Rainwater Harvesting Systems.

Contemporary system designs require an analysis of not only the economic and technical performance of a system, but also the environmental performance. Life Cycle Assessment is a methodology used to evaluate the environmental impacts of a precut or systems, from cradle-to-grave of its' lifetime. Devkota et al., developed such a methodology for rainwater harvesting, and found that the building design (e.g., dimensions) and function (e.g., educational, residential, etc.) play critical roles in the environmental performance of the system. The Economic and Environmental Analysis of Sanitations Technologies, EEAST model evaluates the greenhouse gas emissions and cost of such systems over the lifetime of a variety of building types.

To address the functional parameters of rainwater harvesting systems, a new metric was developed - the demand to supply ratio (D/S) - identifying the ideal building design (supply) and function (demand) in regard to the environmental performance of rainwater harvesting for toilet flushing. With the idea that supply of rainwater not only saves the potable water, but also saves the stormwater entering the combined sewer network (thereby requiring treatment), the savings in environmental emissions were higher if the buildings are connected to a combined sewer network compared to separate one.

Rain Water Harvesting by Freshwater Flooded Forests

Ratagul Freshwater Flooded Forest, Bangladesh

Rain water harvesting is possible by growing fresh water flooded forests without losing the income from the used /submerged land. The main purpose of the rain water harvesting is to utilize the locally available rain water to meet water requirements throughout the year without the need of huge capital expenditure. This would facilitate availability of uncontaminated water for domestic, industrial and irrigation needs.

New Approaches

Instead of using the roof for catchment, the RainSaucer, which looks like an upside down umbrella, collects rain straight from the sky. This decreases the potential for contamination and makes potable water for developing countries a potential application. Other applications of this free standing rainwater collection approach are sustainable gardening and small plot farming.

A Dutch invention called the Groasis Waterboxx is also useful for growing trees with harvested and stored dew and rainwater.

Presentation of RainSaucer system to students at Orphanage in Guatemala.

Traditionally, storm water management using detention basins served a single purpose. However, Optimized Real-Time Control (OptiRTC) lets this infrastructure double as a source of rainwater harvesting without compromising the existing detention capacity. This has been used in the EPA headquarters to evacuate stored water prior to storm events, thus reducing wet weather flow while ensuring water availability for later reuse. This has the benefit of increasing water quality released and decreasing the volume of water released during combined sewer overflow events.

Generally, check dams are constructed across the streams to enhance the percolation of surface water in to the sub soil strata. The water percolation in the water impounded area of the check dams, can be enhanced artificially many folds by loosening the sub soil strata / overburden by using ANFO explosives as used in open cast mining. Thus local aquifers can be recharged quickly by using the available surface water fully for using in the dry season.

History

Earlier

Around the third century BC, the farming communities in Balochistan (now located in Pakistan, Afghanistan and Iran), and Kutch, India, used rainwater harvesting for agriculture and many uses also. In ancient Tamil Nadu , rainwater harvesting was done by Chola kings. Rainwater from the Brihadeeswarar temple (located in Balaganpathy Nagar, Thanjavur, India) was collected in Shivaganga tank. During the later Chola period, the Vīrānam tank was built (1011 to 1037 CE) in Cuddalore district of Tamil Nadu to store water for drinking and irrigation purposes. Vīrānam is a 16 km (9.9 mi) long tank with a storage capacity of 1,465,000,000 cubic feet (41,500,000 m³).

Rainwater harvesting was done in the Indian states of Madhya Pradesh, Maharashtra, and Chhattisgarh in the olden days. Ratanpur, in the state of Chhattisgarh, had around 150 ponds. Most of the tanks or ponds were utilized in agriculture works.

It is a little know fact that the town of Venice depended for centuries on rainwater harvesting.

The lagoon which surrounds Venice is made of brackish water which is not suitable for human drinking. The ancient inhabitants of Venice established a system of rainwater collection which was based on man-made insulated collection wells. Water would percolate down the specially designed stone flooring, and be filtered by a layer of sand, then collecting at the bottom of the well. Later, as Venice acquired territories on the mainland, it started to import water by boat from local rivers, but the wells remained in use, and were especially important in time of war when access to the mainland water could be blocked by an assailant.

Current use

- In China and Brazil rooftop rainwater harvesting is being practiced for providing drinking water, domestic water, water for livestock, water for small irrigation and a way to replenish groundwater levels. Gansu province in China and semi-arid north east Brazil have the largest rooftop rainwater harvesting projects ongoing.

- In Bermuda, the law requires all new construction to include rainwater harvesting adequate for the residents.

- The U.S. Virgin Islands have a similar law.

- In Senegal and Guinea-Bissau, the houses of the Diola-people are frequently equipped with homebrew rainwater harvesters made from local, organic materials.

- In the Irrawaddy Delta of Myanmar, the groundwater is saline and communities rely on mud-lined rainwater ponds to meet their drinking water needs throughout the dry season. Some of these ponds are centuries old and are treated with great reverence and respect.

- In the United States: until 2009 in Colorado, water rights laws almost completely restricted rainwater harvesting; a property owner who captured rainwater was deemed to be stealing it from those who have rights to take water from the watershed. Now, residential well owners that meet certain criteria may obtain a permit to install a *rooftop precipitation collection system* (SB 09-080). Up to 10 large scale pilot studies may also be permitted (HB 09-1129). The main factor in persuading the Colorado Legislature to change the law was a 2007 study that found that in an average year, 97% of the precipitation that fell in Douglas County, in the southern suburbs of Denver, never reached a stream—it was used by plants or evaporated on the ground. In Colorado you cannot even drill a water well unless you have at least 35 acres (14 ha). In New Mexico, rainwater catchment is mandatory for new dwellings in Santa Fe. Texas offers a sales tax exemption on the purchase of rainwater harvesting equipment. Both Texas and Ohio allow the practice even for potable purposes. Oklahoma passed the Water for 2060 Act in 2012, to promote pilot projects for rainwater and graywater use among other water saving techniques.

- In Beijing, some housing societies are now adding rain water in their main water sources after proper treatment.

- In Ireland, Professor Micheal Mcginley established a project to design a rain water harvesting prototype in the Biosystems design Challenge Module at University College Dublin

Canada

A number of Canadians have started implementing rainwater harvesting systems for use in storm-water reduction, irrigation, laundry, and lavatory plumbing. Substantial reform to Canadian law since the mid 2000s has increased the use of this technology in agricultural, industrial, and residential use; but ambiguity remains amongst legislation in many provinces. Bylaws and local municipal codes often regulate rainwater harvesting.

India

- Karnataka: In Bangalore it is mandatory for adoption of rain water harvesting for every owner or the occupier of a building having the sital area measuring 60 ft (18.3 m) X 40 ft (12.2 m) and above and for newly constructed building measuring 30 ft (9.1 m) X 40 ft (12.2 m) and above dimension. In this regard BWSSB has initiated and constructed "Rain Water Harvesting Theme Park" in the name of Sir. M. Visvesvaraya in 1.2 acres (4,900 m²) land situated at Jayanagar, Bangalore. In this park 26 different type of Rain Water Harvesting models are demonstrated along with the water conservation tips. The Auditorium on the first floor is set up with Green Air conditioning system and the same will be utilized to arrange the meeting and showing the video clip about the rain water harvesting to students as well as general public.

- Tamil Nadu: In the state of Tamil Nadu, rainwater harvesting was made compulsory for every building to avoid groundwater depletion. It gave excellent results within five years, and every state took it as role model. Since its implementation, Chennai saw a 50 percent rise in water level in five years and the water quality significantly improved.

- Rajasthan: In Rajasthan, rainwater harvesting has traditionally been practiced by the people of the Thar Desert. There are many ancient water harvesting systems in Rajasthan, which have now been revived. Water harvesting systems are widely used in other areas of Rajasthan as well, for example the chauka system from the Jaipur district.

- Kerala:

- Maharashtra: At present, in Pune, rainwater harvesting is compulsory for any new housing society to be registered.

- In Mumbai city in Maharashtra rain water harvesting is being considered as a good solution to solve water crisis.

The Mumbai city council is planning to make rainwater harvesting mandatory for large societies. An attempt has been made at the Department of Chemical Engineering, IISc, Bangalore to harvest rainwater using upper surface of a solar still, which was used for water distillation

Israel

The Southwest Center for the Study of Hospital and Healthcare Systems in cooperation with Rotary International is sponsoring a rainwater harvesting model program across the country. The first rainwater catchment system was installed at an elementary school in Lod, Israel. The project is looking to expand to Haifa in its third phase. The Southwest Center has also partnered with the Water Resources Action Project (WRAP) of Washington D.C. WRAP currently has rainwater harvesting projects in the West Bank. Rainwater harvesting systems are being installed in local schools for the purpose of educating schoolchildren about water conservation principles and bridging divides between people of different religious and ethnic backgrounds all while addressing the water scarcity issue that the Middle East faces.

New Zealand

Although New Zealand has plentiful rainfall in the West and South, for much of the country, rain water harvesting is the normal practice for most rural housing and is encouraged by most councils

Sri Lanka

Rainwater harvesting has been a popular method of obtaining water for agriculture and for drinking purposes in rural homes. The legislation to promote rainwater harvesting was enacted through the Urban Development Authority (Amendment) Act, No. 36 of 2007. Lanka rainwater harvesting forum is leading the Sri Lanka's initiative.

South Africa

The South African Water Research Commission has supported research into rainwater harvesting. Reports on this research are available on their 'Knowledge Hub'. Studies in arid, semi-arid and humid regions have confirmed that techniques such as mulching, pitting, ridging and modified run-on plots are effective for small-scale crop production.

United Kingdom

In the United Kingdom, water butts are often found in domestic gardens and on allotments to collect rainwater, which is then used to water the garden. However, the British government's Code For Sustainable Homes encouraged fitting large underground tanks to new-build homes to collect rainwater for flushing toilets, watering and washing. Ideal designs had the potential to reduce demand on mains water supply by half. The code was revoked in 2015.

United States

- Colorado: Until 2009, water rights laws almost completely restricted rainwater harvesting; a property owner who captured rainwater was deemed to be stealing it from those who have rights to take water from the watershed. Now, residential well owners that meet certain criteria may obtain a permit to install a *rooftop precipitation collection system* (SB 09-080). Up to 10 large scale pilot studies may also be permitted (HB 09-1129). The main factor in persuading the Colorado Legislature to change the law was a 2007 study that found that in an average year, 97% of the precipitation that fell in Douglas County, in the southern suburbs of Denver, never reached a stream—it was used by plants or evaporated on the ground. In Colorado you cannot even drill a water well unless you have at least 35 acres (14 ha)

- New Mexico: In New Mexico, rainwater catchment is mandatory for new dwellings in Santa Fe.

- Texas offers a sales tax exemption on the purchase of rainwater harvesting equipment. Both Texas and Ohio allow the practice even for potable purposes.

- Oklahoma passed the Water for 2060 Act in 2012, to promote pilot projects for rainwater and graywater use among other water saving techniques.

Non-traditional

- In 1992, American artist Michael Jones McKean created an artwork in Omaha, Nebraska at the Bemis Center for Contemporary Art that created a fully sustainable rainbow in the Omaha skyline. The project collected thousands of gallons of rainwater storing the water in six daisy-chained 12,000 gallons tanks. The massive logistical undertaking, during its 5-month span, was one of the largest urban rainwater harvesting sites in the American Mid-West.

Reclaimed Water

Samples of different types of (waste)water, starting with raw sewage then plant effluent and finally reclaimed water (after several treatment steps)

Reclaimed water or recycled water, is former wastewater (sewage) that is treated to remove solids and impurities, and used in sustainable landscaping irrigation, to recharge groundwater aquifers, to meet commercial and industrial water needs, and for drinking. The purpose of these processes is water conservation and sustainability, rather than discharging the treated water to surface waters such as rivers and oceans. In some cases, recycled water can be used for streamflow augmentation to benefit ecosystems and improve aesthetics. One example of this is along Calera Creek in the City of Pacifica, CA.

The definition of reclaimed water, as defined by Levine and Asano, is "The end product of waste-water reclamation that meets water quality requirements for biodegradable materials, suspended matter and pathogens." Simply stated, reclaimed water is water that is used more than one time before it passes back into the natural water cycle. Scientifically-proven advances in water technology allow communities to reuse water for many different purposes, including industrial, irrigation, and drinking. The water is treated differently depending upon the source and use of the water and how it gets delivered.

Cycled repeatedly through the planetary hydrosphere, all water on Earth is recycled water, but the terms "recycled water" or "reclaimed water" typically mean wastewater sent from a home or business through a pipeline system to a treatment facility, where it is treated to a level consistent with its intended use. The water is then routed directly to a recycled water system for uses such as irrigation or industrial cooling.

There are examples of communities that have safely used recycled water for many years. Los Angeles County's sanitation districts have provided treated wastewater for landscape irrigation in parks and golf courses since 1929. The first reclaimed water facility in California was built at San Francisco's Golden Gate Park in 1932. The Irvine Ranch Water District (IRWD) was the first water district in California to receive an unrestricted use permit from the state for its recycled water; such a permit means that water can be used for any purpose except drinking. IRWD maintains one of the largest recycled water systems in the nation with more than 400 miles serving more than 4,500 metered connections. The Irvine Ranch Water District and Orange County Water District in Southern California are established leaders in recycled water. Further, the Orange County Water District, located in Orange County, and in other locations throughout the world such as Singapore, water is given more advanced treatments and is used indirectly for drinking.

In spite of quite simple methods that incorporate the principles of water-sensitive urban design (WSUD) for easy recovery of stormwater runoff, there remains a common perception that reclaimed water must involve sophisticated and technically complex treatment systems, attempting to recover the most complex and degraded types of sewage. As this effort is driven by sustainability factors, this type of implementation should inherently be associated with point source solutions, where it is most economical to achieve the expected outcomes. Harvesting of stormwater or rainwater can be an extremely simple to comparatively complex, as well as energy and chemical intensive, recovery of more contaminated sewage.

Terminology

Effluent storage tank from where treated effluent (after constructed wetland) is pumped away for irrigation, Haran-Al-Awamied, Syria

There is no one-size-fits-all solution to water reuse, but there are many safe and scientifically-proven options that allow communities to sustain their local water supplies. Below are terms scientists and water experts use to describe some of these reclaimed water options:

Reused water is water used more than once or recycled.

Potable water is drinking water.

Potable reuse refers to reused water you can drink.

Nonpotable reuse refers to reused water that is not used for drinking, but is safe to use for irrigation or industrial purposes.

De facto, unacknowledged or unplanned potable reuse occurs when water intakes draw raw water supplies downstream from discharges of treated effluent from wastewater treatment plants/water reclamation facilities or resource recovery facilities. For example, if you are downstream of a community, that community's used water (run-off and treated wastewater) gets put back into river or stream and is delivered downstream to your community and becomes part of your drinking water supply.

Planned potable reuse is publicly acknowledged as an intentional project to recycle water for drinking water. It can be either direct or indirect. It commonly involves a more formal public process and public consultation program than is observed with de facto or unacknowledged reuse.

How potable reused water is delivered determines if it is called Indirect Potable Reuse or Direct Potable Reuse.

- Indirect potable reuse means the water is delivered to you indirectly. After it is purified, the reused water blends with other supplies and/or sits a while in some sort of storage, man-made or natural, before it gets delivered to a pipeline that leads to a water treatment plant or distribution system. That storage could be a groundwater basin or a surface water reservoir.

- Direct potable reuse means the reused water is put directly into pipelines that go to a water treatment plant or distribution system. Direct potable reuse may occur with or without "engineered storage" such as underground or above ground tanks.

Greywater uses the same waste as water reclamation with the exception of toilet water. Greywater can be used for irrigation, toilet flushing or other domestic uses.

Desalination is an energy-intensive process where salt and other minerals are removed from sea water to produce potable water for drinking and irrigation, typically through membrane filtration (reverse-osmosis), and steam-distillation.

Maximum water recovery - To determine maximum water recovery there are various techniques that have been developed by researchers; for maximum water reuse/reclamation/recovery strategies such as water pinch analysis. The techniques help a user to target the minimum freshwater consumption and wastewater target. It also helps in designing the network that achieves the target. This provides a benchmark to be used by users in improving their water systems.

Applications

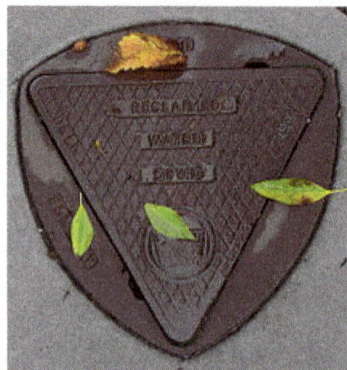

Cover for reclaimed water valve, San Francisco Water District

Most of the uses of water reclamation are non potable uses such as: Washing Cars, flushing toilets, cooling water for power plants, concrete mixing, artificial lakes, irrigation for golf courses and public parks, and for hydraulic fracturing. Where applicable, systems run a dual piping system to keep the recycled water separate from the potable water.

New technologies for recycling allow the water to be used for fracking purposes and can save an estimated 4 - 7 million or more gallons per well.

Potable uses

Some water agencies reuse highly treated effluent from municipal wastewater or resource recovery plants as a reliable, drought proof source of drinking water. By using advanced purification processes, they produce water that meets all applicable drinking water standards. System reliability and frequent monitoring and testing are imperative to them meeting stringent controls.

The water needs of a community, water sources, public health regulations, costs, and the types of water infrastructure in place, such as distribution systems, man-made reservoirs, or natural groundwater basins, determine if and how reclaimed water can be part of the drinking water supply. Communities in El Paso, Texas and Orange County, California, for example, reuse water to replenish groundwater basins. Others, such as the Upper Occoquan Service Authority in Virginia, put it into surface water reservoirs. In these instances the reclaimed water is blended with other water supplies and/or sits in storage for a certain amount of time before it is drawn out and gets treated again at a water treatment or distribution system. In some Texas communities, the reused water is put directly into pipelines that go to a water treatment plant or distribution system. In Singapore reclaimed water is called NEWater and is bottled directly from an advanced water purification facility for educational and celebratory purposes. Though most of the reused water is used for high-tech industry in Singapore, a small amount is returned to reservoirs for drinking water.

A 2012 study conducted by the National Research Council in the United States of America found that the risk of exposure to certain microbial and chemical contaminants from drinking reclaimed water does not appear to be any higher than the risk experienced in at least some current drinking water treatment systems, and may be orders of magnitude lower. This report recommends adjustments to the federal regulatory framework that could enhance public health protection for both planned and unplanned (or de facto) reuse and increase public confidence in water reuse.

Modern technologies such as reverse osmosis and ultraviolet disinfection are commonly used when reclaimed water will be mixed with the drinking water supply. An experiment by the University of New South Wales reportedly showed a reverse osmosis system removed ethinylestradiol and paracetamol from the wastewater, even at 1000 times the expected concentration.

Aboard the International Space Station, astronauts have been able to drink recycled urine due to the introduction of the ECLSS system. The system costs $250 million and has been working since May 2009. The system recycles wastewater and urine back into potable water used for drinking, food preparation, and oxygen generation. This cuts back on the need for resupplying the space station so often.

Indirect Potable Reuse

Some municipalities are using and others are investigating Indirect Potable Reuse (IPR) of reclaimed water. For example, reclaimed water may be pumped into (subsurface recharge) or percolated down to (surface recharge) groundwater aquifers, pumped out, treated again, and finally used as drinking water. This technique may also be referred to as *groundwater recharging*. This includes slow processes of further multiple purification steps via the layers of earth/sand (absorption) and microflora in the soil (biodegradation).

Direct Potable Reuse

In a Direct Potable Reuse (DPR) scheme, water is put directly into pipelines that go to a water treatment plant or distribution system. Direct potable reuse may occur with or without "engineered storage" such as underground or above ground tanks. Communities in Texas have implemented DPR projects, and the state of California is studying the feasibility of developing DPR regulations.

Unplanned Potable Reuse

Water reuse occurs in various ways throughout the world. It happens daily on rivers and other water bodies everywhere. If you live in a community downstream of another, chances are you are reusing its water and likewise communities downstream of you are most likely reusing your water. Unplanned Indirect Potable Use has existed even before the introduction of reclaimed water. Many cities already use water from rivers that contain effluent discharged from upstream sewage treatment plants. There are many large towns on the River Thames upstream of London (Oxford, Reading, Swindon, Bracknell) that discharge their treated sewage ("non-potable water") into the river, which is used to supply London with water downstream.

This phenomenon is also observed in the United States, where the Mississippi River serves as both the destination of sewage treatment plant effluent and the source of potable water. Research conducted in the 1960s by the London Metropolitan Water Board demonstrated that the maximum extent of recycling water is about 11 times before the taste of water induces nausea in sensitive individuals. This is caused by the buildup of inorganic ions such as Cl^-, SO_4^{2-}, K^+ and Na^+, which are not removed by conventional sewage treatment.

Space Travel

Wastewater reclamation can be especially important in relation to human spaceflight.

- In 1998, NASA announced it had built a human waste reclamation bioreactor designed for use in the International Space Station and a manned Mars mission. Human urine and feces are input into one end of the reactor and pure oxygen, pure water, and compost (humanure) are output from the other end. The soil could be used for growing vegetables, and the bioreactor also produces electricity.

Design Considerations

Distribution

Nonpotable reclaimed water is often distributed with a dual piping network that keeps reclaimed water pipes completely separate from potable water pipes. In the United States and some other countries, nonpotable reclaimed water is distributed in lavender (light purple) pipes to distinguish it from potable water. The use of the color purple for pipes carrying recycled water was pioneered by the Irvine Ranch Water District in Irvine, California.

In many cities using reclaimed water, it is now in such demand that consumers are only allowed to use it on assigned days. Some cities that previously offered unlimited reclaimed water at a flat rate are now beginning to charge citizens by the amount they use.

Reclamation Processes

Wastewater must pass through numerous systems before being returned to the environment. Here is a partial listing from one particular plant system:

- Barscreens - Barscreens remove large solids that are sent into a grinder. All solids are then dumped into a sewer pipe at a Treatment Plant.

- Primary Settling Tanks - Readily settable and floatable solids are removed from the wastewater. These solids are skimmed from the top and bottom of the tanks and sent to the Treatment Plant where it'll be turned into fertilizer.

- Biological Treatment - The wastewater is cleaned through a biological treatment method that uses microorganisms, bacteria which digest the sludge and reduce the nutrient content. Air bubbles up to keep the organisms suspended and to supply oxygen to the aerobic bacteria so they can metabolize the food, convert it to energy, CO_2, and water, and reproduce more microorganisms. This helps to remove ammonia also through nitrification.

- Secondary Settling Tanks - The force of the flow slows down as sewage enters these tanks, allowing the microorganisms to settle to the bottom. As they settle, other small particles suspended in the water are picked up, leaving behind clear wastewater. Some of the microorganisms that settle to the bottom are returned to the system to be used again.

- Tertiary Treatment - Deep-bed, single-media, gravity sand filters receive water from the secondary basins and filter out the remaining solids. As this is the final process to remove solids, the water in these filters is almost completely clear.

- Chlorine Contact Tanks - Three chlorine contact tanks disinfect the water to decrease the risks associated with discharging wastewater containing human pathogens. This step protects the quality of the waters that receive the wastewater discharge.

One of two procedures are then followed according to the future disposal site:

1. Reclaimed Water Pump Station - The pump station distributes reclaimed water to users around the City. This may include golf courses, agricultural uses, cooling towers, or in land fills.

2. Water is passed through high level purification to be returned to the environment. Currently this means a reverse osmosis system.

Treatment Improvements

As world populations require both more clean water and better ways to dispose of wastewater, the demand for water reclamation will increase. Future success in water reuse will depend on whether this can be done without adverse effects on human health and the environment.

In the United States, reclaimed waste water is generally treated to secondary level when used for irrigation, but there are questions about the adequacy of that treatment. Some leading scientists in the main water society, AWWA, have long believed that secondary treatment is insufficient to protect people against pathogens, and recommend adding at least membrane filtration, reverse osmosis, ozonation, or other advanced treatments for irrigation water.

There have been recent advances in reverse osmosis in different countries, but have consistently produced very high quality water all the same. In Singapore, reclaimed water, also known as NE-Water has become cleaner than the government issue tap water. Also, according to Bartels, the Bedok Demonstration Plant, which uses RO membranes, has successfully run for the past 3 years, producing high quality wastewater all the while.

Seepage of nitrogen and phosphorus into ground and surface water is also becoming a serious problem, and will probably lead to at least tertiary treatment of reclaimed water to remove nutrients in the future. Even using secondary treatment, water quality can be improved. Water quality can also be improved as it passes through the subsurface mixing zone where surface water and groundwater combine. Testing for pathogens using Polymerase Chain Reaction (PCR) instead of older culturing techniques, and changing the discredited fecal coloform "indicator organism" standard would be improvements.

In a large study treatment plants showed that they could significantly reduce the numbers of parasites in effluent, just by making adjustments to the currently used process. But, even using the best of current technology, risk of spreading drug resistance in the environment through wastewater effluent, would remain.

Some scientists have suggested that there need to be basic changes in treatment, such as using bacteria to degrade waste based on nitrogen (urine) and not just carbonaceous (fecal) waste, saying that this would greatly improve effectiveness of treatment. Currently designed plants do not deal well with contaminants in solution (e.g. pharmaceuticals). "Dewatering" solids is a major problem. Some wastes could be disposed of without mixing them with water to begin with. In an interesting innovation, solids (sludge) could be removed before entering digesters and burned into a gas that could be used to run engines.

Emerging disinfection technologies include ultrasound, pulse arc electrohydraulic discharge, and bank filtration. Another issue is concern about weakened mandates for pretreatment of industrial wastes before they are made part of the municipal waste stream. Some also believe that hospitals should treat their own wastes. The safety of drinking reclaimed water which has been given advanced treatment and blended with other waters, remains controversial.

In recent years, as hydraulic fracturing of oil and gas formations has become more and more commonplace, new technologies for water recycling have emerged. One such technology uses a combination of ozone and electrocoagulation. This process removes organics, hydrocarbons, spent polymers, chemical additives used in the fracturing process, and heavy metals such as barium, iron, boron and more.

Alternatives

Seawater Desalination

In urban areas where climate change has threatened long-term water security and reduced rainfall over catchment areas, using reclaimed water for indirect potable use may be superior to other water supply augmentation methods. One other commonly used option is seawater desalination. Recycling wastewater and desalinating seawater may have many of the same disadvantages, including high costs of water treatment, infrastructure construction, transportation, and waste disposal problems. Although the best option varies from region to region, desalination is often superior economically, as reclaimed water usually requires a dual piping network, often with additional storage tanks, when used for nonpotable use.

Greywater Systems

A less elaborate alternative to reclaimed water is a greywater system. Greywater is wastewater that has been used in sinks, baths, showers, or washing machines, but does not contain sewage and has not been treated at the same levels as recycled water. In a home system, treat-ed or untreated greywater may be used to flush toilets or for irrigation. Some systems now exist which directly use greywater from a sink to flush a toilet.

Rainwater Harvesting

Perhaps the simplest option is a rainwater harvesting system. Although there are concerns about the quality of rainwater in urban areas, due to air pollution and acid rain, many systems exist now to use untreated rainwater for nonpotable uses or treated rainwater for direct potable use. Urban design systems which incorporate rainwater harvesting and reduce runoff are known as Water Sensitive Urban Design (WSUD) in Australia, Low Impact Development (LID) in the United States and Sustainable urban drainage systems (SUDS) in the United Kingdom. There are also concerns about rainwater harvesting systems reducing the amount of run-off entering natural bodies of water.

Health Aspects

Reclaimed water is highly engineered for safety and reliability so that the quality of reclaimed water is more predictable than many existing surface and groundwater sources. Reclaimed water is considered safe when appropriately used. Reclaimed water planned for use in recharging aquifers or augmenting surface water receives adequate and reliable treatment before mixing with naturally occurring water and undergoing natural restoration processes. Some of this water eventually becomes part of drinking water supplies.

A water quality study published in 2009 compared the water quality differences of reclaimed/recycled water, surface water, and groundwater. Results indicate that reclaimed water, surface water, and groundwater are more similar than dissimilar with regard to constituents. The researchers tested for 244 representative constituents typically found in water. When detected, most constituents were in the parts per billion and parts per trillion range. DEET (a bug repellant), and Caffeine were found in all water types and virtually in all samples. Triclosan (in anti-bacterial soap & toothpaste) was found in all water types, but detected in higher levels (parts per trillion) in reclaimed water than in surface or groundwater. Very few hormones/steroids were detected in samples, and when detected were at very low levels. Haloacetic acids (a disinfection by-product) were found in all types of samples, even groundwater. The largest difference between reclaimed water and the other waters appears to be that reclaimed water has been disinfected and thus has disinfection by-products (due to chlorine use).

A 2005 study titled "Irrigation of Parks, Playgrounds, and Schoolyards with Reclaimed Water" found that there had been no incidences of illness or disease from either microbial pathogens or chemicals, and the risks of using reclaimed water for irrigation are not measurably different from irrigation using potable water. Studies by the National Academies of Science, the Monterey Regional Water Pollution Control Agency, and others have found reclaimed water to be safe for agricultural use.

Testing Standards

Reclaimed water is not regulated by the Environmental Protection Agency (EPA), but the EPA has developed water reuse guidelines that were most recently updated in 2012. The EPA Guidelines for Water Reuse represents the international standard for best practices in water reuse. The document was developed under a Cooperative Research and Development Agreement between the U.S. Environmental Protection Agency (EPA), the U.S. Agency for International Development (USAID), and the global consultancy CDM Smith. The Guidelines provide a framework for states to develop regulations that incorporate the best practices and address local requirements.

Ongoing wastewater research sometimes raise concerns about pathogens in the water. Many pathogens cannot be detected by currently used tests.

Recent literature also questions the validity of testing for "indicator organisms" instead of pathogens. Nor do present standards consider interactions of heavy metals and pharmaceuticals which may foster the development of drug resistant pathogens in waters derived from sewage.

To address these concerns about the source water, reclaimed water providers use multi-barrier treatment processes and constant monitoring to ensure that reclaimed water is safe and treated properly for the intended end use.

Potable use

The main health risk for potable use of reclaimed water is the potential for pharmaceutical and other household chemicals or their derivatives (Environmental persistent pharmaceutical pollutants) to persist in this water. This would be of much less concern if the population were to keep their excrement out of the wastewater e.g. via the use of the Urine-diverting dry toilet or systems that treat blackwater separately from greywater.

Environmental Aspects

There is debate about possible health and environmental effects. To address these concerns, A

Risk Assessment Study of potential health risks of recycled water and comparisons to conventional Pharmaceuticals and Personal Care Product (PPCP) exposures was conducted by the WateReuse Research Foundation. For each of four scenarios in which people come into contact with recycled water used for irrigation - children on the playground, golfers, and landscape, and agricultural workers - the findings from the study indicate that it could take anywhere from a few years to millions of years of exposure to nonpotable recycled water to reach the same exposure to PPCPs that we get in a single day through routine activities.

Using reclaimed water for non-potable uses saves potable water for drinking, since less potable water will be used for non-potable uses.

It sometimes contains higher levels of nutrients such as nitrogen, phosphorus and oxygen which may somewhat help fertilize garden and agricultural plants when used for irrigation.

The usage of water reclamation decreases the pollution sent to sensitive environments. It can also enhance wetlands, which benefits the wildlife depending on that eco-system. It also helps to stop the chances of drought as recycling of water reduces the use of fresh water supply from underground sources. For instance, The San Jose/Santa Clara Water Pollution Control Plant instituted a water recycling program to protect the San Francisco Bay area's natural salt water marshes.

Costs and Evaluation

The cost of reclaimed water exceeds that of potable water in many regions of the world, where a fresh water supply is plentiful. However, reclaimed water is usually sold to citizens at a cheaper rate to encourage its use. As fresh water supplies become limited from distribution costs, increased population demands, or climate change reducing sources, the cost ratios will evolve also. The evaluation of reclaimed water needs to consider the entire water supply system, as it may bring important value of flexibility into the overall system

History

Storm and sanitary sewers were necessarily developed along with the growth of cities. By the 1840s the luxury of indoor plumbing, which mixes human waste with water and flushes it away, eliminated the need for cesspools. Odor was considered the big problem in waste disposal and to address it, sewage could be drained to a lagoon, or "settled" and the solids removed, to be disposed of separately. This process is now called "primary treatment" and the settled solids are called "sludge."

At the end of the 19th century, since primary treatment still left odor problems, it was discovered that bad odors could be prevented by introducing oxygen into the decomposing sewage. This was the beginning of the biological aerobic and anaerobic treatments which are fundamental to waste water processes.

By the 1920s, it became necessary to further control the pollution caused by the large quantities of human and industrial liquid wastes which were being piped into rivers and oceans, and modern treatment plants were being built in the US and other industrialized nations by the 1930s.

Designed to make water safe for fishing and recreation, the Clean Water Act of 1972 mandated elimination of the discharge of untreated waste from municipal and industrial sources, and the

US federal government provided billions of dollars in grants for building sewage treatment plants around the country. Modern treatment plants, usually using oxidation and/or chlorination in addition to primary and secondary treatment, were required to meet certain standards.

Current treatment improves the quality of separated wastewater solids or sludge. The separated water is given further treatment considered adequate for non potable use by local agencies, and discharged into bodies of water, or reused as reclaimed water. In places like Florida, where it is necessary to avoid nutrient overload of sensitive receiving water, reuse of treated or reclaimed water can be more economically feasible than meeting the higher standards for surface water disposal mandated by the Clean Water Act

Examples

Indirect Potable Reuse (IPR)

- Orange County, California

- Pasadena, California

- Singapore (where it is branded as *NEWater*)

- Payson, Arizona

- The Torreele project in the Veurne coastal region of Belgium, which began operating in 2002

- Virginia Occoquan Reservoir - The Upper Occoquan Sewage Authority plant discharges its highly treated output to supply roughly 20% of the inflow into the Occoquan Reservoir, which provides drinking water used by the Fairfax County Water Authority - one of the three major water providers in the Washington, D.C. metropolitan area.

Non-potable Reuse (NPR)

- Austin, Texas

- Caboolture and Maroochy (South East Queensland, Australia) LGA's currently provide reclaimed water for industrial use (primarily capital works). Users must apply for a key to be able to access the compounds in which the outlets are located.

- Clark County, Nevada

- Clearwater, Florida

- Contra Costa County, California

- Melbourne, Australia

- Mount Buller Ski resort uses recycled water for snow making.

- San Antonio operates the largest recycled water system in the United States.

- Sydney, Australia

- Tucson, Arizona

- San Diego, California (San Diego County)

- St. Petersburg, Florida

Proposed

In some places, reclaimed water has been proposed for either potable or non-potable use:

- South East Queensland, Australia (planned for potable use as of late 2010)

- Newcastle, New South Wales, Australia (proposed for non-potable use).

- Canberra, Australian Capital Territory, Australia (proposed in January 2007 as a backup source of potable water)

- Los Angeles, California - By 2019, the Los Angeles Department of Water and Power will build a plant to replenish their groundwater aquifer with purified water in order to deal with the shortage of rain and snow fall, restricted water imports and local groundwater contamination.

Israel

As of 2010, Israel leads the world in the proportion of water it recycles. Israel treats 80% of its sewage (400 billion liters a year), and 100% of the sewage from the Tel Aviv metropolitan area is treated and reused as irrigation water for agriculture and public works. The remaining sludge is currently pumped into the Mediterranean, however a new bill has passed stating a conversion to treating the sludge to be used as manure. Only 20% of the treated water is lost (due to evaporation, leaks, overflows and seeping). The recycled water allows farmers to plan ahead and not be limited by water shortages. There are many levels of treatment, and many different ways of treating the water—which leads to a big difference in the quality of the end product. The best quality of reclaimed sewage water comes from adding a gravitational filtering step, after the chemical and biological cleansing. This method uses small ponds in which the water seeps through the sand into the aquifer in about 400 days, then is pumped out as clear purified water. This is nearly the same process used in the space station water recycling system, which turns urine and feces into purified drinking water, oxygen and manure.

To add to the efficiency of the Israeli system - the reclaimed sewage water may be mixed with reclaimed sea water (Plans are in action to increase the desalinization program up to 50% of the countries usage by 2013 - 600 billion liters of drinkable sea water a year), along with aquifer water and fresh sweet lake water - monitored by computer to account for the nationwide needs and input. This action reduced the outdated risk of salt and mineral percentages in the water. Plans to implement this overall usage of reclaimed water for drinking are discouraged by the psychological preconception of the public for the quality of reclaimed water, and the fear of its origin. As of today, all the reclaimed sewage water in Israel is used for agricultural and land improvement purposes.

U.S.

The leaders in use of reclaimed water in the U.S. are Florida and California, with Irvine Ranch

Water District as one of the leading developers. They were the first district to approve the use of reclaimed water for in-building piping and use in flushing toilets.

In a January 2012 U.S. National Research Council report, a committee of independent experts found that expanding the reuse of municipal wastewater for irrigation, industrial uses, and drinking water augmentation could significantly increase the United States' total available water resources. The committee noted that a portfolio of treatment options is available to mitigate water quality issues in reclaimed water. The report also includes a risk analysis that suggests the risk of exposure to certain microbial and chemical contaminants from drinking reclaimed water is not any higher than the risk from drinking water from current water treatment systems—and in some cases, may be orders of magnitude lower. The report concludes that adjustments to the federal regulatory framework could enhance public health protection and increase public confidence in water reuse.

Australia

As Australia continues to battle the 7–10-year drought, nationwide, reclaimed effluent is becoming a popular option. Two major capital cities in Australia, Adelaide and Brisbane, have already committed to adding reclaimed effluent to their dwindling dams. The former has also built a desalination plant to help battle any future water shortages. Brisbane has been seen as a leader in this trend, and other cities and towns will review the Western Corridor Recycled Water Project once completed. Goulbourn, Canberra, Newcastle, and Regional Victoria, Australia are already considering building a reclaimed effluent process.

European Union

The second largest waste reclamation program in the world is in Spain, where 12% of the nation's waste is treated.

According to an EU-funded study, "Europe and the Mediterranean countries are lagging behind" California, Japan, and Australia "in the extent to which reuse is being taken up." According to the study "the concept (of reuse) is difficult for the regulators and wider public to understand and accept."

Water Metering

Water metering is the process of measuring water use.

In many developed countries water meters are used to measure the volume of water used by residential and commercial building that are supplied with water by a public water supply system. Water meters can also be used at the water source, well, or throughout a water system to determine flow through a particular portion of the system. In most of the world water meters measure flow in cubic metres (m^3) or litres but in the USA and some other countries water meters are calibrated in cubic feet ($ft.^3$) or US gallons on a mechanical or electronic register. Some electronic meter registers can display rate-of-flow in addition to total usage.

A typical residential water meter

There are several types of water meters in common use. The choice depends on the flow measurement method, the type of end user, the required flow rates, and accuracy requirements.

In North America, standards for manufacturing water meters are set by the American Water Works Association.

Types of Metering Devices

There are two common approaches to flow measurement, displacement and velocity, each making use of a variety of technologies. Common displacement designs include oscillating piston and nutating disc meters. Velocity-based designs include single- and multi-jet meters and turbine meters.

There are also non-mechanical designs, for example electromagnetic and ultrasonic meters, and meters designed for special uses. Most meters in a typical water distribution system are designed to measure cold potable water only. Specialty hot water meters are designed with materials that can withstand higher temperatures. Meters for reclaimed water have special lavender register covers to signify that the water should not be used for drinking.

Additionally, there are electromechanical meters, like prepaid water meters and automatic meter reading meters. The latter integrates an electronic measurement component and a LCD with a mechanical water meter. Mechanical water meters normally use a reed switch, hall or photoelectric coding register as the signal output. After processing by the microcontroller unit (MCU) in the electronic module, the data are transmitted to the LCD or output to an information management system.

Water meters are generally owned, read and maintained by a public water provider such as a city, rural water association or private water company. In some cases an owner of a mobile home park, apartment complex or commercial building may be billed by a utility based on the reading of one meter, with the costs shared among the tenants based on some sort of key (size of flat, number of inhabitants or by separately tracking the water consumption of each unit in what is called submetering).

Displacement Water Meters

Water meter in Belo Horizonte.

This type of water meter is most often used in residential and small commercial applications and homes. Displacement meters are commonly referred to as Positive Displacement, or "PD" meters. Two common types are oscillating piston meters and nutating disk meters. Either method relies on the water to physically displace the moving measuring element in direct proportion to the amount of water that passes through the meter. The piston or disk moves a magnet that drives the register.

PD meters are generally very accurate at the low-to-moderate flow rates typical of residential and small commercial users, and commonly range in size from 5/8" to 2". Because displacement meters require that all water flows through the meter to "push" the measuring element, they generally are not practical in large commercial applications requiring high flow rates or low pressure loss. PD meters normally have a built-in strainer to protect the measuring element from rocks or other debris that could stop or break the measuring element. PD meters normally have bronze, brass or plastic bodies with internal measuring chambers made of molded plastics and stainless steel.

Velocity Water Meters

A velocity-type meter measures the velocity of flow through a meter of a known internal capacity. The speed of the flow can then be converted into volume of flow to determine the usage. There are several types of meters that measure water flow velocity, including jet meters (single-jet and multi-jet), turbine meters, propeller meters and mag meters. Most velocity-based meters have an adjustment vane for calibrating the meter to the required accuracy.

Multi-jet Meters

Multi-jet meters are very accurate in small sizes and are commonly used in 5/8" to 2" sizes for residential and small commercial users. Multi-jet meters use multiple ports surrounding an internal chamber to create multiple jets of water against an impeller, whose rotation speed depends on the velocity of water flow. Multi-jets are very accurate at low flow rates, but there are no large size meters since they do not have the straight-through flow path needed for the high flow rates used in large pipe diameters. Multi-jet meters generally have an internal strainer element that can protect the jet ports from getting clogged. Multi-jet meters normally have bronze alloy bodies or outer casings, with internal measuring parts made from modern thermoplastics and stainless steel.

Turbine Meters

Turbine meters are less accurate than displacement and jet meters at low flow rates, but the measuring element does not occupy or severely restrict the entire path of flow. The flow direction is generally straight through the meter, allowing for higher flow rates and less pressure loss than displacement-type meters. They are the meter of choice for large commercial users, fire protection and as master meters for the water distribution system. Strainers are generally required to be installed in front of the meter to protect the measuring element from gravel or other debris that could enter the water distribution system. Turbine meters are generally available for 1-½" to 12" or higher pipe sizes. Turbine meter bodies are commonly made of bronze, cast iron or ductile iron. Internal turbine elements can be plastic or non-corrosive metal alloys. They are accurate in normal working conditions but are greatly affected by the flow profile and fluid conditions.

- Fire meters are a specialized type of turbine meter meeting the high flow rates requirements for fire protection. They are often approved by Underwriters Laboratories (UL) or Factory Mutual (FM) for use in fire protection.

- Fire hydrant meters are a specialized type of portable turbine meter attached to a fire hydrant to measure water flowing out of the hydrant. The meters are normally made of aluminium to keep their weight low and usually have a 3" capacity. Utilities often require them for measuring water used on construction sites, for pool filling, or where a permanent meter has not yet been installed.

Compound Meters

A compound meter is used where high flow rates are necessary, but where at times there are also smaller rates of flow that need to be accurately measured. Compound meters have two measuring elements and a check valve to regulate flow between them. At high flow rates, water is normally diverted primarily or completely to the high flow element. The high flow element is typically a turbine meter. When flow rates drop to where the high flow element cannot measure accurately, a check valve closes to divert water to a smaller element that can measure the lower flow rates accurately. The low flow element is typically a multi-jet or PD meter. By adding the values registered by the high and low elements, the utility has a record of the total consumption of water flowing through the meter.

Electromagnetic Meters

Magnetic flow meters, commonly referred to as "mag meters", are technically a velocity-type water meter, except that they use electromagnetic properties to determine the water flow velocity, rather than the mechanical means used by jet and turbine meters. Mag meters use the physics principle of Faraday's law of induction for measurement, and require AC or DC electricity from a power line or battery to operate the electromagnets. Since mag meters have no mechanical measuring element, they normally have the advantage of being able to measure flow in either direction, and use electronics for measuring and totalizing the flow. Mag meters can also be useful for measuring raw (untreated/unfiltered) water and waste-water, since there is no mechanical measuring element to get clogged or damaged by debris flowing through the meter. Strainers are not required with mag meters, since there is no measuring element in the stream of flow that could be damaged. Since stray electrical energy flowing through the flow tube can cause inaccurate readings, most mag meters are installed with either grounding rings or grounding electrodes to divert stray electricity away from the electrodes used to measure the flow inside the flow tube.

Electromagnetic flow meter

Ultrasonic Meters

Ultrasonic water meters use an ultrasonic transducer to send ultrasonic sound waves through the fluid to determine the velocity of the water. Since the cross-sectional area of the meter body is a fixed and known value, when the velocity of water is detected, the volume of water passing through the meter can be calculated with very high accuracy. Because water density changes with temperature, most ultrasonic water meters also measure the water temperature as a component of the volume calculation.

There are 2 primary ultrasonic measurement technologies used in water metering:

- Doppler effect meters utilize the Doppler Effect to determine the velocity of water passing through the meter.

- Transit Time meters measure the amount of time required for the ultrasonic signal to pass between 2 or more fixed points inside the meter.

Ultrasonic meters may either be of flow-through or "clamp-on" design. Flow through designs are those where the water passes directly through the meter, and are typically found in residential or commercial applications. Clamp-on designs are generally used for larger diameters where the sensors are mounted to the exterior of pipes, etc.

Ultrasonic water meters are typically very accurate, with residential meters capable of measuring down to 0.01 gallons or 0.001 cubic feet. In addition they have wide flow measurement ranges, require little maintenance and have long lifespans due to the lack of internal mechanical components to wear out. While relatively new to the American water utility market, ultrasonic meters have been used in commercial applications for many years and are becoming widely accepted due to their advantages over traditional mechanical designs.

Prepaid Water Meters

Meters can be prepaid or postpaid, depending on the payment method. Most mechanical type water meters are of the postpaid type, as are electromagnetic and ultrasonic meters. With prepaid water meters the user purchases and prepays for a given amount of water from a vending station. The amount of water credited is entered on media such as an IC or RF type card. The main difference is whether the card needs a contact with the processing part of the prepaid water meter. In some areas a prepaid water meter uses a keypad as the interface for inputting the water credit.

Registers

A typical water meter register showing a meter reading of 8.3 gallons. Notice the black "1" on the odometer has not yet fully turned over, so only the red hand is read.

Water meters connected to remote reading devices through three-wire cables

There are several types of registers on water meters. A standard register normally has a dial similar to a clock, with gradations around the perimeter to indicate the measuring unit and the amount of water used, if less than the lowest digit in a display similar to the odometer wheels in a car, their sum being the total volume used. Modern registers are normally driven by a magnetic coupling between a magnet in the measuring chamber attached to the measuring element and another attached to the bottom of the register. Gears in the register convert the motion of the measuring element to the proper usage increment for display on the sweep hand and the odometer-style wheels. Many registers also have a leak detector. This is a small visible disk or hand that is geared closer to the rotation speed of the drive magnet, so that very small flows that would be visually undetectable on the regular sweep hand can be seen.

With Automatic Meter Reading, manufacturers have developed pulse or encoder registers to produce electronic output for radio transmitters, reading storage devices, and data logging devices. Pulse meters send a digital or analog electronic pulse to a recording device. Encoder registers have an electronic means permitting an external device to interrogate the register to obtain either the position of the wheels or a stored electronic reading. Frequent transmissions of consumption data can be used to give smart meter functionality.

There are also some specialized types of registers such as meters with an LCD instead of mechanical wheels, and registers to output data or pulses to a variety of recording and controller devices. For industrial applications, output is often 4-20 mA analog for recording or controlling different flow rates in addition to totalization.

Water Meter Reading

Different size meters indicate different resolutions of the reading. One rotation of the sweep hand may be equivalent to 10 gallons or to 1,000 gallons (1 to 100 ft.^{3}, 0.1 to 10 m^3). If one rotation of the hand represents 10 gallons, the meter has a 10-gallon sweep. Sometimes the last number(s) of the wheel display are non-rotating or printed on the dial face. The fixed zero number(s) are represented by the position of the rotating sweep hand. For example, if one rotation of the hand is 10 gallons, the sweep hand is on 7, and the wheel display shows 123456 plus a fixed zero, the actual total usage would be 1,234,567 gallons.

In the United States most utilities bill only to the nearest 100 or 1,000 gallons (10 to 100 ft.^{3}, 1 to 10 m^3), and often only read the leftmost 4 or 5 numbers on the display wheels. Using the above example, they would read and bill 1,234, rounding to 1,234,000 gallons based on a 1,000 gallon billing resolution. The most common rounding for a particular size meter is often indicated by differently coloured number wheels, the ones ignored being black, and the ones used for billing being white.

Prevalence

Water metering is common for residential and commercial drinking water supply in many countries, as well as for industrial self-supply with water. However, it is less common in irrigated agriculture, which is the major water user worldwide. Water metering is also uncommon for piped drinking water supply in rural areas and small towns, although there are examples of successful metering in rural areas in developing countries, such as in El Salvador.

Metering of water supplied by utilities to residential, commercial and industrial users is common in most developed countries, except for the United Kingdom where only about 38% of users are metered. In some developing countries metering is very common, such as in Chile where it stands at 96%, while in others it still remains low, such as in Argentina.

The percentage of residential water metering in selected cities in developing countries is as follows:

- 99% in Santiago de Chile (1998)

- 96% in Abidjan, Ivory Coast (1987)

- 62% in cities in Guatemala (2000)

- 30% in Lima, Peru (1991)

- 28% in Kathmandu, Nepal (2001)

- 2% in Buenos Aires, Argentina (1992)

Nearly two-thirds of OECD countries meter more than 90% of single-family houses. A few are also expanding their metering of apartments (e.g., France and Germany).

Benefits

The benefits of metering are that:

- in conjunction with volumetric pricing it provides an incentive for water conservation,

- it helps to detect water leaks in the distribution network, thus providing a basis for reducing the amount of non-revenue water;

- it is a precondition for quantity-targeting of water subsidies to the poor.

Costs

The costs of metering include:

- Investment costs to purchase, install and replace meters,

- Recurring costs to read meters and issue bills based on consumption instead of bills based on monthly flat fees.

While the cost of purchasing residential meters is low, the total life cycle costs of metering are high. For example, retrofitting flats in large buildings with meters for every flat can involve major and thus costly plumbing work.

Problems

Problems associated with metering arise particularly in the case of intermittent supply, which is common in many developing countries. Sudden changes in pressure can damage meters to the extent that many meters in cities in developing countries are not functional. Also, some types of meters become less accurate as they age, and under-registering consumption leads to lower revenues if defective meters are not regularly replaced. Many types of meters also register air flows, which can lead to over-registration of consumption, especially in systems with intermittent supply, when water supply is re-established and the incoming water pushes air through the meters.

Effect on Consumption

There is disagreement as to the effect of metering and water pricing on water consumption. The price elasticity of metered water demand varies greatly depending on local conditions. The effect of volumetric water pricing on consumption tends to be higher if the water bill represents a significant portion of household expenditures. There is evidence from the UK that there is an instant drop in consumption of some 10% when meters are installed. In Hamburg, Germany, domestic water consumption for metered flats (112 liter/capita/day) was 18% lower than for unmetered flats (137 liter/capita/day) in 1992. The municipal utility Hamburger Wasserwerke GmbH had installed more than 40,000 water meters in individual flats of older houses since 1985. All new apartments had to be metered by law. Previously there had been only a single meter for the entire house in multi-apartment houses.

References

- Martin V. Melosi (2010). The Sanitary City: Environmental Services in Urban America from Colonial Times to the Present. University of Pittsburgh Press. p. 110. ISBN 9780822973379.

- Colin A. Russell (2003). Edward Frankland: Chemistry, Controversy and Conspiracy in Victorian England. Cambridge University Press. pp. 372–380. ISBN 9780521545815.

- Sharma, Sanjay Kumar; Sanghi, Rashmi (2012). Advances in Water Treatment and Pollution Prevention. Springer. ISBN 9789400742048. Retrieved 2013-02-07.

- Tilley, David F. (2011). Aerobic Wastewater Treatment Processes: History and Development. IWA Publishing. ISBN 9781843395423. Retrieved 2013-02-07.

- Water Reuse: Potential for Expanding the Nation's Water Supply through Reuse of Municipal Wastewater. National Research Council. 2012. ISBN 978-0-309-25749-7.

- Helgeson, Tom (2009). A Reconnaissance-Level Quantitative Comparison of Reclaimed Water, Surface Water, and Groundwater. Alexandria, VA: WateReuse Research Foundation. p. 141. ISBN 978-1-934183-12-0.

- Crook, James (2005). Irrigation of Parks, Playgrounds, and Schoolyards: Extent and Safety. Alexandria, VA: WateReuse Research Foundation. p. 60. ISBN 0-9747586-3-9.

- Tchobanoglous, George; Burton, Franklin L.; Stensel, H. David; Metcalf & Eddy, Inc. (2003). Wastewater Engineering: Treatment and Reuse (4th ed.). McGraw-Hill. ISBN 0-07-112250-8.

- "Water Recycling and Reuse: The Environmental Benefits/". US Environment Protection Agency. 23 February 2016. Retrieved 22 August 2016.

- Mund, Jan-Peter. "Capacities for Megacities coping with water scarcity" (PDF). UN-Water Decade Programme on Capacity Development. Retrieved 2014-02-17.

- Ashton, John; Ubido, Janet (1991). "The Healthy City and the Ecological Idea" (PDF). Journal of the Society for the Social History of Medicine. 4 (1): 173–181. doi:10.1093/shm/4.1.173. Retrieved 8 July 2013.

- Kumar, Ro. "Collect up to 10 gallons of water per inch of rain with Rainsaucers' latest standalone rainwater catchment". LocalBlu. Retrieved 11 February 2013.

- "82(R) H.B. No. 3391. Act relating to rainwater harvesting and other water conservation initiatives. † went into effect on September 1, 2011". 82nd Regular Session. Texas Legislature Online. Retrieved 8 February 2013.

- "82(R) H.B. No. 3391. Act relating to rainwater harvesting and other water conservation initiatives. † went into effect on September 1, 2011". 82nd Regular Session. Texas Legislature Online. Retrieved 8 February 2013.

- Devkota, Jay (2013). "Development and application of EEAST: A life cycle based model for use of harvested rainwater and composting toilets in buildings". Journal of Environmental Management.

- Ong, Hian Hai; Ryck, Luc De, "Best Sourcing Approach Keeps Water Production Costs Down", Water & Wastewater International: 13, retrieved 22 March 2012

Water Conservation Techniques in Agriculture

This chapter discusses the techniques used for water conservation in agriculture. The techniques discussed within this chapter are irrigation, surface irrigation, dip irrigation and other approaches of water conservation. Due to environmental concerns and the depletion of fresh water resources, it is very important to further develop and practice water conservation.

Irrigation

Irrigation is the method in which water is supplied to plants at regular intervals for agriculture. It is used to assist in the growing of agricultural crops, maintenance of landscapes, and revegetation of disturbed soils in dry areas and during periods of inadequate rainfall. Additionally, irrigation also has a few other uses in crop production, which include protecting plants against frost, suppressing weed growth in grain fields and preventing soil consolidation. In contrast, agriculture that relies only on direct rainfall is referred to as rain-fed or dry land farming.

An irrigation sprinkler watering a lawn

Irrigation systems are also used for dust suppression, disposal of sewage, and in mining. Irrigation is often studied together with drainage, which is the natural or artificial removal of surface and sub-surface water from a given area.

Irrigation has been a central feature of agriculture for over 5,000 years and is the product of many cultures. Historically, it was the basis for economies and societies across the globe, from Asia to the Southwestern United States.

Irrigation canal in Osmaniye, Turkey

History

Animal-powered irrigation, Upper Egypt, ca. 1846

An example of an irrigation system common on the Indian subcontinent. Artistic impression
on the banks of Dal Lake, Kashmir, India

Inside a karez tunnel at Turpan, Sinkiang

Archaeological investigation has identified as evidence of irrigation where the natural rainfall was insufficient to support crops for rainfed agriculture.

Perennial irrigation was practiced in the Mesopotamian plain whereby crops were regularly watered throughout the growing season by coaxing water through a matrix of small channels formed in the field.

irrigation in Tamil Nadu (India)

Ancient Egyptians practiced *Basin irrigation* using the flooding of the Nile to inundate land plots which had been surrounded by dykes. The flood water was held until the fertile sediment had settled before the surplus was returned to the watercourse. There is evidence of the ancient Egyptian pharaoh Amenemhet III in the twelfth dynasty (about 1800 BCE) using the natural lake of the Faiyum Oasis as a reservoir to store surpluses of water for use during the dry seasons, the lake swelled annually from flooding of the Nile.

The Ancient Nubians developed a form of irrigation by using a waterwheel-like device called a *sakia*. Irrigation began in Nubia some time between the third and second millennium BCE. It largely depended upon the flood waters that would flow through the Nile River and other rivers in what is now the Sudan.

In sub-Saharan Africa irrigation reached the Niger River region cultures and civilizations by the first or second millennium BCE and was based on wet season flooding and water harvesting.

Terrace irrigation is evidenced in pre-Columbian America, early Syria, India, and China. In the Zana Valley of the Andes Mountains in Peru, archaeologists found remains of three irrigation canals radiocarbon dated from the 4th millennium BCE, the 3rd millennium BCE and the 9th century CE. These canals are the earliest record of irrigation in the New World. Traces of a canal possibly dating from the 5th millennium BCE were found under the 4th millennium canal. Sophisticated irrigation and storage systems were developed by the Indus Valley Civilization in present-day Pakistan and North India, including the reservoirs at Girnar in 3000 BCE and an early canal irrigation system from circa 2600 BCE. Large scale agriculture was practiced and an extensive network of canals was used for the purpose of irrigation.

Ancient Persia (modern day Iran) as far back as the 6th millennium BCE, where barley was grown in areas where the natural rainfall was insufficient to support such a crop. The Qanats, developed in ancient Persia in about 800 BCE, are among the oldest known irrigation methods still in use today. They are now found in Asia, the Middle East and North Africa. The system comprises a network of vertical wells and gently sloping tunnels driven into the sides of cliffs and steep hills to tap groundwater. The noria, a water wheel with clay pots around the rim powered by the flow of the stream (or by animals where the water source was still), was first brought into use at about this time, by Roman settlers in North Africa. By 150 BCE the pots were fitted with valves to allow smoother filling as they were forced into the water.

The irrigation works of ancient Sri Lanka, the earliest dating from about 300 BCE, in the reign of King Pandukabhaya and under continuous development for the next thousand years, were one of the most complex irrigation systems of the ancient world. In addition to underground canals, the Sinhalese were the first to build completely artificial reservoirs to store water. Due to their engineering superiority in this sector, they were often called 'masters of irrigation'. Most of these irrigation systems still exist undamaged up to now, in Anuradhapura and Polonnaruwa, because of the advanced and precise engineering. The system was extensively restored and further extended during the reign of King Parakrama Bahu (1153–1186 CE).

China

The oldest known hydraulic engineers of China were Sunshu Ao (6th century BCE) of the Spring and Autumn Period and Ximen Bao (5th century BCE) of the Warring States period, both of whom worked on large irrigation projects. In the Sichuan region belonging to the State of Qin of ancient China, the Dujiangyan Irrigation System was built in 256 BCE to irrigate an enormous area of farmland that today still supplies water. By the 2nd century AD, during the Han Dynasty, the Chinese also used chain pumps that lifted water from lower elevation to higher elevation. These were powered by manual foot pedal, hydraulic waterwheels, or rotating mechanical wheels pulled by oxen. The water was used for public works of providing water for urban residential quarters and palace gardens, but mostly for irrigation of farmland canals and channels in the fields.

Korea

In 15th century Korea, the world's first rain gauge, *uryanggye* (Korean:우량계), was invented in 1441. The inventor was Jang Yeong-sil, a Korean engineer of the Joseon Dynasty, under the active direction of the king, Sejong the Great. It was installed in irrigation tanks as part of a nationwide system to measure and collect rainfall for agricultural applications. With this instrument, planners and farmers could make better use of the information gathered in the survey.

North America

In North America, the Hohokam were the only culture to rely on irrigation canals to water their crops, and their irrigation systems supported the largest population in the Southwest by AD 1300. The Hohokam constructed an assortment of simple canals combined with weirs in their various agricultural pursuits. Between the 7th and 14th centuries, they also built and maintained extensive irrigation networks along the lower Salt and middle Gila rivers that rivaled the complexity of those used in the ancient Near East, Egypt, and China. These were constructed using relatively simple excavation tools, without the benefit of advanced engineering technologies, and achieved drops of a few feet per mile, balancing erosion and siltation. The Hohokam cultivated varieties of cotton, tobacco, maize, beans and squash, as well as harvested an assortment of wild plants. Late in the Hohokam Chronological Sequence, they also used extensive dry-farming systems, primarily to grow agave for food and fiber. Their reliance on agricultural strategies based on canal irrigation, vital in their less than hospitable desert environment and arid climate, provided the basis for the aggregation of rural populations into stable urban centers.

Present Extent

Irrigation ditch in Montour County, Pennsylvania, off Strawberry Ridge Road

In the mid-20th century, the advent of diesel and electric motors led to systems that could pump groundwater out of major aquifers faster than drainage basins could refill them. This can lead to permanent loss of aquifer capacity, decreased water quality, ground subsidence, and other problems. The future of food production in such areas as the North China Plain, the Punjab, and the Great Plains of the US is threatened by this phenomenon.

At the global scale, 2,788,000 km² (689 million acres) of fertile land was equipped with irrigation infrastructure around the year 2000. About 68% of the area equipped for irrigation is located in Asia, 17% in the America, 9% in Europe, 5% in Africa and 1% in Oceania. The largest contiguous areas of high irrigation density are found:

- In Northern India and Pakistan along the Ganges and Indus rivers

- In the Hai He, Huang He and Yangtze basins in China

- Along the Nile river in Egypt and Sudan

- In the Mississippi-Missouri river basin and in parts of California

Smaller irrigation areas are spread across almost all populated parts of the world.

Only eight years later in 2008, the scale of irrigated land increased to an estimated total of 3,245,566 km² (802 million acres), which is nearly the size of India.

Types of Irrigation

Basin flood irrigation of wheat

Irrigation of land in Punjab, Pakistan

Various types of irrigation techniques differ in how the water obtained from the source is distributed within the field. In general, the goal is to supply the entire field uniformly with water, so that each plant has the amount of water it needs, neither too much nor too little.

Surface Irrigation

In *surface* (*furrow*, *flood*, or *level basin*) irrigation systems, water moves across the surface of agricultural lands, in order to wet it and infiltrate into the soil. Surface irrigation can be subdivided into furrow, *borderstrip or basin irrigation*. It is often called *flood irrigation* when the irrigation results in flooding or near flooding of the cultivated land. Historically, this has been the most common method of irrigating agricultural land and still is in most parts of the world.

Where water levels from the irrigation source permit, the levels are controlled by dikes, usually plugged by soil. This is often seen in terraced rice fields (rice paddies), where the method is used to flood or control the level of water in each distinct field. In some cases, the water is pumped, or lifted by human or animal power to the level of the land. The field water efficiency of surface irrigation is typically lower than other forms of irrigation but has the potential for efficiencies in the range of 70% - 90% under appropriate management.

Localized Irrigation

Impact sprinkler head

Localized irrigation is a system where water is distributed under low pressure through a piped network, in a pre-determined pattern, and applied as a small discharge to each plant or adjacent to it. Drip irrigation, spray or micro-sprinkler irrigation and bubbler irrigation belong to this category of irrigation methods.

Subsurface Textile Irrigation

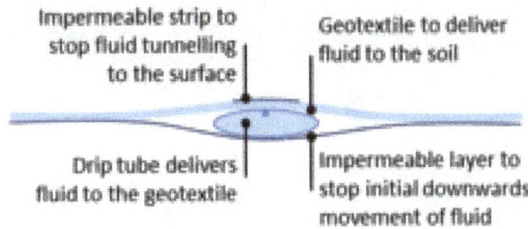

Impermeable strip to stop fluid tunnelling to the surface

Geotextile to deliver fluid to the soil

Drip tube delivers fluid to the geotextile

Impermeable layer to stop initial downwards movement of fluid

Diagram showing the structure of an example SSTI installation

Subsurface Textile Irrigation (SSTI) is a technology designed specifically for subsurface irrigation in all soil textures from desert sands to heavy clays. A typical subsurface textile irrigation system has an impermeable base layer (usually polyethylene or polypropylene), a drip line running along that base, a layer of geotextile on top of the drip line and, finally, a narrow impermeable layer on top of the geotextile. Unlike standard drip irrigation, the spacing of emitters in the drip pipe is not critical as the geotextile moves the water along the fabric up to 2 m from the dripper.

Drip Irrigation

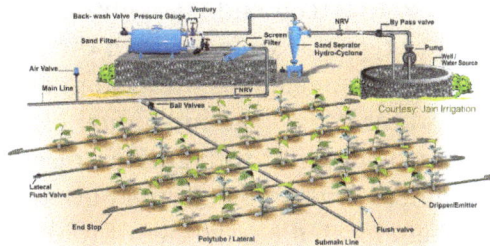

Drip irrigation layout and its parts

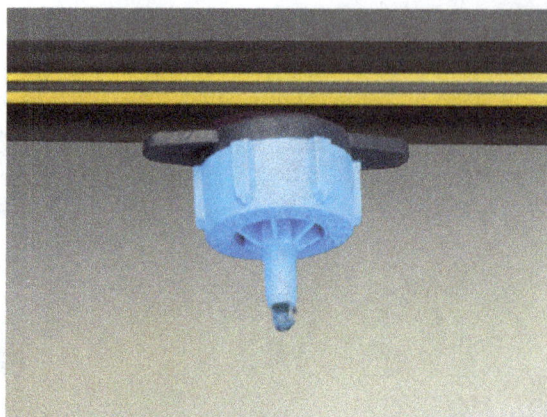

Drip irrigation – a dripper in action

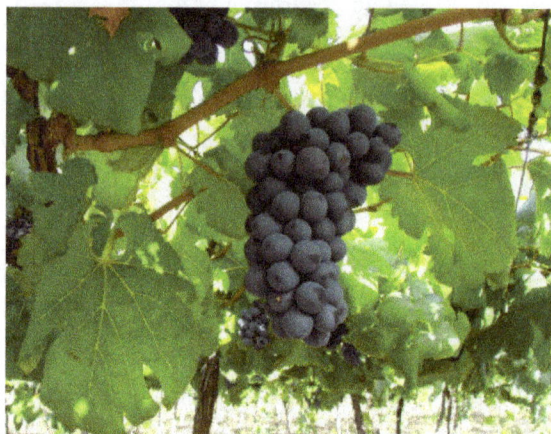

Grapes in Petrolina, only made possible in this semi arid area by drip irrigation

Drip (or micro) irrigation, also known as trickle irrigation, functions as its name suggests. In this system water falls drop by drop just at the position of roots. Water is delivered at or near the root zone of plants, drop by drop. This method can be the most water-efficient method of irrigation, if managed properly, since evaporation and runoff are minimized. The field water efficiency of drip irrigation is typically in the range of 80 to 90 percent when managed correctly.

In modern agriculture, drip irrigation is often combined with plastic mulch, further reducing evaporation, and is also the means of delivery of fertilizer. The process is known as *fertigation*.

Deep percolation, where water moves below the root zone, can occur if a drip system is operated for too long or if the delivery rate is too high. Drip irrigation methods range from very high-tech and computerized to low-tech and labor-intensive. Lower water pressures are usually needed than for most other types of systems, with the exception of low energy center pivot systems and surface irrigation systems, and the system can be designed for uniformity throughout a field or for precise water delivery to individual plants in a landscape containing a mix of plant species. Although it is difficult to regulate pressure on steep slopes, pressure compensating emitters are available, so the field does not have to be level. High-tech solutions involve precisely calibrated emitters located along lines of tubing that extend from a computerized set of valves.

Irrigation using Sprinkler Systems

Sprinkler irrigation of blueberries in Plainville, New York, United States

A traveling sprinkler at Millets Farm Centre, Oxfordshire, United Kingdom

In *sprinkler* or overhead irrigation, water is piped to one or more central locations within the field and distributed by overhead high-pressure sprinklers or guns. A system utilizing sprinklers, sprays, or guns mounted overhead on permanently installed risers is often referred to as a *solid-set* irrigation system. Higher pressure sprinklers that rotate are called *rotors* an are driven by a ball drive, gear drive, or impact mechanism. Rotors can be designed to rotate in a full or partial circle. Guns are similar to rotors, except that they generally operate at very high pressures of 40 to 130 lbf/in² (275 to 900 kPa) and flows of 50 to 1200 US gal/min (3 to 76 L/s), usually with nozzle diameters in the range of 0.5 to 1.9 inches (10 to 50 mm). Guns are used not only for irrigation, but also for industrial applications such as dust suppression and logging.

Sprinklers can also be mounted on moving platforms connected to the water source by a hose. Automatically moving wheeled systems known as *traveling sprinklers* may irrigate areas such as small farms, sports fields, parks, pastures, and cemeteries unattended. Most of these utilize a length of polyethylene tubing wound on a steel drum. As the tubing is wound on the drum powered by the irrigation water or a small gas engine, the sprinkler is pulled across the field. When the sprinkler arrives back at the reel the system shuts off. This type of system is known to most people as a "waterreel" traveling irrigation sprinkler and they are used extensively for dust suppression, irrigation, and land application of waste water.

Other travelers use a flat rubber hose that is dragged along behind while the sprinkler platform is pulled by a cable. These cable-type travelers are definitely old technology and their use is limited in today's modern irrigation projects.

Center pivot with drop sprinklers

Wheel line irrigation system in Idaho, 2001

Center pivot irrigation

Center pivot irrigation is a form of sprinkler irrigation consisting of several segments of pipe (usually galvanized steel or aluminium) joined together and supported by trusses, mounted on wheeled towers with sprinklers positioned along its length. The system moves in a circular pattern and is fed with water from the pivot point at the center of the arc. These systems are found and used in all parts of the world and allow irrigation of all types of terrain. Newer systems have drop sprinkler heads as shown in the image that follows.

Most center pivot systems now have drops hanging from a u-shaped pipe attached at the top of the pipe with sprinkler head that are positioned a few feet (at most) above the crop, thus limiting evaporative losses. Drops can also be used with drag hoses or bubblers that deposit the water directly on the ground between crops. Crops are often planted in a circle to conform to the center pivot. This type of system is known as LEPA (Low Energy Precision Application). Originally, most center pivots were water powered. These were replaced by hydraulic systems (*T-L Irrigation*) and electric motor driven systems (Reinke, Valley, Zimmatic). Many modern pivots feature GPS devices.

Irrigation by Lateral Move (Side Roll, Wheel Line, Wheelmove)

A *series of pipes, each with a wheel* of about 1.5 m diameter permanently affixed to its midpoint, and sprinklers along its length, are coupled together. Water is supplied at one end using a large hose. After sufficient irrigation has been applied to one strip of the field, the hose is removed, the water drained from the system, and the assembly rolled either by hand or with a purpose-built mechanism, so that the sprinklers are moved to a different position across the field. The hose is reconnected. The process is repeated in a pattern until the whole field has been irrigated.

This system is less expensive to install than a center pivot, but much more labor-intensive to operate - it does not travel automatically across the field: it applies water in a stationary strip, must be drained, and then rolled to a new strip. Most systems use 4 or 5-inch (130 mm) diameter aluminum pipe. The pipe doubles both as water transport and as an axle for rotating all the wheels. A drive system (often found near the centre of the wheel line) rotates the clamped-together pipe sections as a single axle, rolling the whole wheel line. Manual adjustment of individual wheel positions may be necessary if the system becomes misaligned.

Wheel line systems are limited in the amount of water they can carry, and limited in the height of crops that can be irrigated. One useful feature of a lateral move system is that it consists of sections that can be easily disconnected, adapting to field shape as the line is moved. They are most often used for small, rectilinear, or oddly-shaped fields, hilly or mountainous regions, or in regions where labor is inexpensive.

Sub-irrigation

Subirrigation has been used for many years in field crops in areas with high water tables. It is a method of artificially raising the water table to allow the soil to be moistened from below the plants' root zone. Often those systems are located on permanent grasslands in lowlands or river valleys and combined with drainage infrastructure. A system of pumping stations, canals, weirs and gates allows it to increase or decrease the water level in a network of ditches and thereby control the water table.

Sub-irrigation is also used in commercial greenhouse production, usually for potted plants. Water is delivered from below, absorbed upwards, and the excess collected for recycling. Typically, a solution of water and nutrients floods a container or flows through a trough for a short period of time, 10–20 minutes, and is then pumped back into a holding tank for reuse. Sub-irrigation in greenhouses requires fairly sophisticated, expensive equipment and management. Advantages are water and nutrient conservation, and labor-saving through lowered system maintenance and automation. It is similar in principle and action to subsurface basin irrigation.

Irrigation Automatically, non-electric using buckets and ropes

Besides the common manual watering by bucket, an automated, natural version of this also exists. Using plain polyester ropes combined with a prepared ground mixture can be used to water plants from a vessel filled with water.

The ground mixture would need to be made depending on the plant itself, yet would mostly consist of black potting soil, vermiculite and perlite. This system would (with certain crops) allow to save expenses as it does not consume any electricity and only little water (unlike sprinklers, water timers, etc.). However, it may only be used with certain crops (probably mostly larger crops that do not need a humid environment; perhaps e.g. paprikas).

Irrigation using Water Condensed from Humid Air

In countries where at night, humid air sweeps the countryside.Water can be obtained from the humid air by condensation onto cold surfaces. This is for example practiced in the vineyards at Lanzarote using stones to condense water or with various fog collectors based on canvas or foil sheets.

In-ground Irrigation

Most commercial and residential irrigation systems are "in ground" systems, which means that everything is buried in the ground. With the pipes, sprinklers, emitters (drippers), and irrigation valves being hidden, it makes for a cleaner, more presentable landscape without garden hoses or other items having to be moved around manually. This does, however, create some drawbacks in the maintenance of a completely buried system.

Most irrigation systems are divided into zones. A zone is a single irrigation valve and one or a group of drippers or sprinklers that are connected by pipes or tubes. Irrigation systems are divided into zones because there is usually not enough pressure and available flow to run sprinklers for an entire yard or sports field at once. Each zone has a solenoid valve on it that is controlled via wire by an irrigation controller. The irrigation controller is either a mechanical (now the "dinosaur" type) or electrical device that signals a zone to turn on at a specific time and keeps it on for a specified amount of time. "Smart Controller" is a recent term for a controller that is capable of adjusting the watering time by itself in response to current environmental conditions. The smart controller determines current conditions by means of historic weather data for the local area, a soil moisture sensor (water potential or water content), rain sensor, or in more sophisticated systems satellite feed weather station, or a combination of these.

When a zone comes on, the water flows through the lateral lines and ultimately ends up at the irrigation emitter (drip) or sprinkler heads. Many sprinklers have pipe thread inlets on the bottom of them which allows a fitting and the pipe to be attached to them. The sprinklers are usually installed with the top of the head flush with the ground surface. When the water is pressurized, the head will pop up out of the ground and water the desired area until the valve closes and shuts off that zone. Once there is no more water pressure in the lateral line, the sprinkler head will retract back into the ground. Emitters are generally laid on the soil surface or buried a few inches to reduce evaporation losses.

Water Sources

Irrigation water can come from groundwater (extracted from springs or by using wells), from surface water (withdrawn from rivers, lakes or reservoirs) or from non-conventional sources like treated wastewater, desalinated water or drainage water. A special form of irrigation using surface water is spate irrigation, also called floodwater harvesting. In case of a flood (spate), water is diverted to normally dry river beds (wadis) using a network of dams, gates and channels and spread over large areas. The moisture stored in the soil will be used thereafter to grow crops. Spate irrigation areas are in particular located in semi-arid or arid, mountainous regions. While flood-water harvesting belongs to the accepted irrigation methods, rainwater harvesting is usually not considered as a form of irrigation. Rainwater harvesting is the collection of runoff water from roofs or unused land and the concentration of this.

Irrigation is underway by pump-enabled extraction directly from the Gumti, seen in the background, in Comilla, Bangladesh.

Around 90% of wastewater produced globally remains untreated, causing widespread water pollution, especially in low-income countries. Increasingly, agriculture uses untreated waste-water as a source of irrigation water. Cities provide lucrative markets for fresh produce, so are attractive to farmers. However, because agriculture has to compete for increasingly scarce water resources with industry and municipal users, there is often no alternative for farmers but to use water polluted with urban waste, including sewage, directly to water their crops. Significant health hazards can result from using water loaded with pathogens in this way, especially if people eat raw vegetables that have been irrigated with the polluted water. The International Water Management Institute has worked in India, Pakistan, Vietnam, Ghana, Ethiopia, Mexico and other countries on various projects aimed at assessing and reducing risks of wastewater irrigation. They advocate a 'multiple-barrier' approach to wastewater use, where farmers are encouraged to adopt various risk-reducing behaviours. These include ceasing irrigation a few days before harvesting to allow pathogens to die off in the sunlight, applying water carefully so it does not contaminate leaves likely to be eaten raw, cleaning vegetables with disinfectant or allowing fecal sludge used in farming to dry before being used as a human manure. The World Health Organization has developed guidelines for safe water use.

There are numerous benefits of using recycled water for irrigation, including the low cost (when compared to other sources, particularly in an urban area), consistency of supply (regardless of season, climatic conditions and associated water restrictions), and general consistency of quality. Irrigation of recycled wastewater is also considered as a means for plant fertilization and particularly nutrient supplementation. This approach carries with it a risk of soil and water pollution through excessive wastewater application. Hence, a detailed understanding of soil water conditions is essential for effective utilization of wastewater for irrigation.

Efficiency

Young engineers restoring and developing the old Mughal irrigation system during
the reign of the Mughal Emperor Bahadur Shah II

Modern irrigation methods are efficient enough to supply the entire field uniformly with water, so that each plant has the amount of water it needs, neither too much nor too little. Water use efficiency in the field can be determined as follows:

- Field Water Efficiency (%) = (Water Transpired by Crop ÷ Water Applied to Field) x 100

Until 1960s, the common perception was that water was an infinite resource. At that time, there were fewer than half the current number of people on the planet. People were not as wealthy as today, consumed fewer calories and ate less meat, so less water was needed to produce their food. They required a third of the volume of water we presently take from rivers. Today, the competition for water resources is much more intense. This is because there are now more than seven billion people on the planet, their consumption of water-thirsty meat and vegetables is rising, and there is increasing competition for water from industry, urbanisation and biofuel crops. To avoid a global water crisis, farmers will have to strive to increase productivity to meet growing demands for food, while industry and cities find ways to use water more efficiently.

Successful agriculture is dependent upon farmers having sufficient access to water. However, water scarcity is already a critical constraint to farming in many parts of the world. With regards to agriculture, the World Bank targets food production and water management as an increasingly global issue that is fostering a growing debate. Physical water scarcity is where there is not enough water to meet all demands, including that needed for ecosystems to function effectively. Arid regions frequently suffer from physical water scarcity. It also occurs where water seems abundant but where resources are over-committed. This can happen where there is overdevelopment of hydraulic infrastructure, usually for irrigation. Symptoms of physical water scarcity include environmental degradation and declining groundwater. Economic scarcity, meanwhile, is caused by a lack of investment in water or insufficient human capacity to satisfy the demand for water. Symptoms of economic water scarcity include a lack of infrastructure, with people often having to fetch water from rivers for domestic and agricultural uses. Some 2.8 billion people currently live in water-scarce areas.

Technical Challenges

Irrigation schemes involve solving numerous engineering and economic problems while minimizing negative environmental impact.

- Competition for surface water rights.

- Overdrafting (depletion) of underground aquifers.

- Ground subsidence (e.g. New Orleans, Louisiana)

- Underirrigation or irrigation giving only just enough water for the plant (e.g. in drip line irrigation) gives poor soil salinity control which leads to increased soil salinity with consequent buildup of toxic salts on soil surface in areas with high evaporation. This requires either leaching to remove these salts and a method of drainage to carry the salts away. When using drip lines, the leaching is best done regularly at certain intervals (with only a slight excess of water), so that the salt is flushed back under the plant's roots.

- Overirrigation because of poor distribution uniformity or management wastes water, chemicals, and may lead to water pollution.

- Deep drainage (from over-irrigation) may result in rising water tables which in some instances will lead to problems of irrigation salinity requiring watertable control by some form of subsurface land drainage.

- Irrigation with saline or high-sodium water may damage soil structure owing to the formation of alkaline soil

- Clogging of filters: It is mostly algae that clog filters, drip installations and nozzles. UV and ultrasonic method can be used for algae control in irrigation systems.

Surface Irrigation

Furrow Irrigation of sugar cane in Australia, 2006

Surface irrigation is defined as the group of application techniques where water is applied and distributed over the soil surface by gravity. It is by far the most common form of irrigation throughout the world and has been practiced in many areas virtually unchanged for thousands of years.

Surface irrigation is often referred to as flood irrigation, implying that the water distribution is uncontrolled and therefore, inherently inefficient. In reality, some of the irrigation practices grouped under this name involve a significant degree of management (for example surge irrigation). Surface irrigation comes in three major types; level basin, furrow and border strip.

Drip irrigation involves enclosed passages, such as pipes and tubing.

Process

The process of surface irrigation can be described using four phases. As water is applied to the top end of the field it will flow or advance over the field length. The advance phase refers to that length of time as water is applied to the top end of the field and flows or advances over the field length. After the water reaches the end of the field it will either run-off or start to pond. The period of time between the end of the advance phase and the shut-off of the inflow is termed the wetting, ponding or storage phase. As the inflow ceases the water will continue to runoff and infiltrate until the entire field is drained. The depletion phase is that short period of time after cut-off when the length of the field is still submerged. The recession phase describes the time period while the water front is retreating towards the downstream end of the field. The depth of water applied to any point in the field is a function of the opportunity time, the length of time for which water is present on the soil surface.

Types of Surface Irrigation

Basin Irrigation

Level basin flood irrigation on wheat

Level basin irrigation has historically been used in small areas having level surfaces that are surrounded by earth banks. The water is applied rapidly to the entire basin and is allowed to infiltrate. In traditional basins no water is permitted to drain from the field once it is irrigated. Basin irrigation is favoured in soils with relatively low infiltration rates(Walker and Skogerboe 1987).This is also a method of surface irrigation. Fields are typically set up to follow the natural contours of the land but the introduction of laser levelling and land grading has permitted the construction of large rectangular basins that are more appropriate for mechanised broadacre cropping.

Drainback Level Basins

Drain back level basins (DBLB) or contour basins are a variant of basin irrigation where the field is divided into a number of terraced rectangular bays which are graded level or have no significant slope. Water is applied to the first bay (usually the highest in elevation) and when the desired depth is applied water is permitted to drain back off that bay and flow to the next bay which is at a lower elevation than the first. Each bay is irrigated in turn using a combination of drainage water from the previous bay and continuing inflow from the supply channel. Successful operation of these systems is reliant on a sufficient elevation drop between successive bays. These systems are commonly used in Australia where rice and wheat are grown in rotation.

Furrow Irrigation

Furrow irrigation system using siphon tubes

Gated pipe supply system

Furrow irrigation is conducted by creating small parallel channels along the field length in the direction of predominant slope. Water is applied to the top end of each furrow and flows down the field under the influence of gravity. Water may be supplied using gated pipe, siphon and head ditch or bankless systems. The speed of water movement is determined by many factors such as slope, surface roughness and furrow shape but most importantly by the inflow rate and soil infiltration rate. The spacing between adjacent furrows is governed by the crop species, common spacings typically range from 0.75 to 2 metres. The crop is planted on the ridge between furrows which may contain a single row of plants or several rows in the case of a bed type system. Furrows may range anywhere from less than 100 m to 2000 m long depending on the soil type, location and crop type. Shorter furrows are commonly associated with higher uniformity of application but result in increasing potential for runoff losses. Furrow irrigation is particularly suited to broad-acre row crops such as cotton, maize and sugar cane. It is also practiced in various horticultural industries such as citrus, stone fruit and tomatoes.

The water can take a considerable period of time to reach the other end, meaning water has been infiltrating for a longer period of time at the top end of the field. This results in poor uniformity with high application at the top end with lower application at the bottom end. In most cases the performance of furrow irrigation can be improved through increasing the speed at which water moves along the field (the advance rate). This can be achieved through increasing flow rates or through the practice of surge irrigation. Increasing the advance rate not only improves the uniformity but also reduces the total volume of water required to complete the irrigation.

Surge Irrigation

Surge Irrigation is a variant of furrow irrigation where the water supply is pulsed on and off in planned time periods (e.g. on for 1 hour off for 1½ hour). The wetting and drying cycles reduce infiltration rates resulting in faster advance rates and higher uniformities than continuous flow. The reduction in infiltration is a result of surface consolidation, filling of cracks and micro pores and the disintegration of soil particles during rapid wetting and consequent surface sealing during each drying phase. On those soils where surging is effective it has been reported to allow completion of the irrigation with a lower overall water usage and therefore higher efficiency and potentially offer the ability to practice deficit irrigation. The effectiveness of surge irrigation is soil type dependent; for example, many clay soils experience a rapid sealing behaviour under continuous flow and therefore surge irrigation offers little benefit.

Bay/Border Strip Irrigation

Border strip, otherwise known as border check or bay irrigation could be considered as a hybrid of level basin and furrow irrigation. The field is divided into a number of bays or strips, each bay is separated by raised earth check banks (borders). The bays are typically longer and narrower compared to basin irrigation and are orientated to align lengthwise with the slope of the field. Typical bay dimensions are between 10-70m wide and 100-700m long. The water is applied to the top end of the bay, which is usually constructed to facilitate free-flowing conditions at the downstream end. One common use of this technique includes the irrigation of pasture for dairy production.

Issues Associated with Surface Irrigation

While surface irrigation can be practiced effectively using the correct management under the right conditions, it is often associated with a number of issues undermining productivity and environmental sustainability:

- Waterlogging - Can cause the plant to shut down delaying further growth until sufficient water drains from the rootzone. Waterlogging may be counteracted by drainage, tile drainage or watertable control by another form of subsurface drainage.

- Deep drainage - Overirrigation may cause water to move below the root zone resulting in rising water tables. In regions with naturally occurring saline soil layers (for example salinity in south eastern Australia) or saline aqifers, these rising water tables may bring salt up into the root zone leading to problems of irrigation salinity.

- Salinization - Depending on water quality irrigation water may add significant volumes of salt to the soil profile. While this is a lesser issue for surface irrigation compared to other irrigation methods (due to the comparatively high leaching fraction), lack of subsurface drainage may restrict the leaching of salts from the soil. This can be remedied by drainage and soil salinity control through flushing.

The aim of modern surface irrigation management is to minimse the risk of these potential adverse impacts.

Drip Irrigation

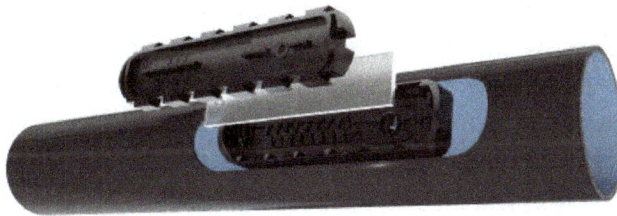

Open pressure compensated dripper

Drip irrigation is a form of irrigation that saves water and fertilizer by allowing water to drip slowly to the roots of many different plants, either onto the soil surface or directly onto the root zone, through a network of valves, pipes, tubing, and emitters. It is done through narrow tubes that deliver water directly to the base of the plant. It is chosen instead of surface irrigation for various reasons, often including concern about minimizing evaporation.

History

Primitive drip irrigation has been used since ancient times. Fan Sheng-Chih Shu, written in China during the first century BCE, describes the use of buried, unglazed clay pots filled with water as a means of irrigation. Modern drip irrigation began its development in Germany in 1860 when researchers began experimenting with subsurface irrigation using clay pipe to create combination irrigation and drainage systems. Research was later expanded in the 1920s to include the application of perforated pipe systems. The usage of plastic to hold and distribute water in drip irrigation was later developed in Australia by Hannis Thill.

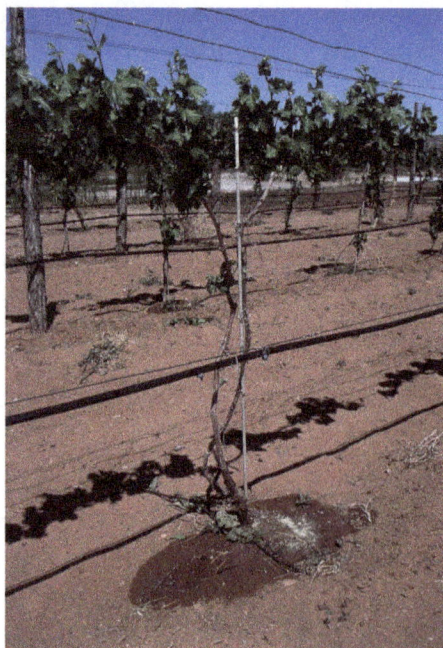

Drip irrigation in Mexico vineyard, 2000

Usage of a plastic emitter in drip irrigation was developed in Israel by Simcha Blass and his son Yeshayahu. Instead of releasing water through tiny holes easily blocked by tiny particles, water was released through larger and longer passageways by using velocity to slow water inside a plastic emitter. The first experimental system of this type was established in 1959 by Blass who partnered later (1964) with Kibbutz Hatzerim to create an irrigation company called Netafim. Together they developed and patented the first practical surface drip irrigation emitter.

In the United States, the first drip tape, called *Dew Hose*, was developed by Richard Chapin of Chapin Watermatics in the early 1960s.

Modern drip irrigation has arguably become the world's most valued innovation in agriculture since the invention of the impact sprinkler in the 1930s, which offered the first practical alternative to surface irrigation. Drip irrigation may also use devices called micro-spray heads, which spray water in a small area, instead of dripping emitters. These are generally used on tree and vine crops with wider root zones. Subsurface drip irrigation (SDI) uses permanently or temporarily buried dripperline or drip tape located at or below the plant roots. It is becoming popular for row crop irrigation, especially in areas where water supplies are limited or recycled water is used for irrigation. Careful study of all the relevant factors like land topography, soil, water, crop and agro-climatic conditions are needed to determine the most suitable drip irrigation system and components to be used in a specific installation.

Components and Operation

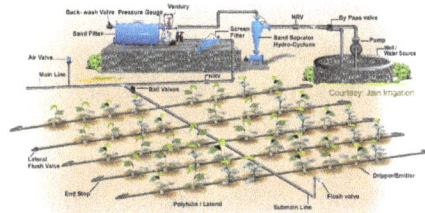

Drip irrigation system layout and its parts

Nursery flowers watered with drip irrigation in Israel

Horticulture drip emitter in a pot

Components used in drip irrigation (listed in order from water source) include:

- Pump or pressurized water source

- Water filter(s) or filtration systems: sand separator, Fertigation systems (Venturi injector) and chemigation equipment (optional)

- Backwash controller (Backflow prevention device)

- Pressure Control Valve (pressure regulator)

- Main line (larger diameter pipe and pipe fittings)

- Hand-operated, electronic, or hydraulic control valves and safety valves

- Smaller diameter polytube (often referred to as "laterals")

- Poly fittings and accessories (to make connections)

- Emitting devices at plants (emitter or dripper, micro spray head, inline dripper or inline driptube)

In drip irrigation systems, pump and valves may be manually or automatically operated by a controller.

Most large drip irrigation systems employ some type of filter to prevent clogging of the small emitter flow path by small waterborne particles. New technologies are now being offered that minimize clogging. Some residential systems are installed without additional filters since potable water is already filtered at the water treatment plant. Virtually all drip irrigation equipment manufacturers recommend that filters be employed and generally will not honor warranties unless this is done. Last line filters just before the final delivery pipe are strongly recommended in addition to any other filtration system due to fine particle settlement and accidental insertion of particles in the intermediate lines.

Drip and subsurface drip irrigation is used almost exclusively when using recycled municipal waste water. Regulations typically do not permit spraying water through the air that has not been fully treated to potable water standards.

Because of the way the water is applied in a drip system, traditional surface applications of timed-release fertilizer are sometimes ineffective, so drip systems often mix liquid fertilizer with the irrigation water. This is called fertigation; fertigation and chemigation (application of pesticides and other chemicals to periodically clean out the system, such as chlorine or sulfuric acid) use chemical injectors such as diaphragm pumps, piston pumps, or aspirators. The chemicals may be added constantly whenever the system is irrigating or at intervals. Fertilizer savings of up to 95% are being reported from recent university field tests using drip fertigation and slow water delivery as compared to timed-release and irrigation by micro spray heads.

Properly designed, installed, and managed, drip irrigation may help achieve water conservation by reducing evaporation and deep drainage when compared to other types of irrigation such as flood or overhead sprinklers since water can be more precisely applied to the plant roots. In addition, drip can eliminate many diseases that are spread through water contact with the foliage. Finally, in regions where water supplies are severely limited, there may be no actual water savings, but rather simply an increase in production while using the same amount of water as before. In very arid regions or on sandy soils, the preferred method is to apply the irrigation water as slowly as possible.

Pulsed irrigation is sometimes used to decrease the amount of water delivered to the plant at any one time, thus reducing runoff or deep percolation. Pulsed systems are typically expensive and require extensive maintenance. Therefore, the latest efforts by emitter manufacturers are focused toward developing new technologies that deliver irrigation water at ultra-low flow rates, i.e. less than 1.0 liter per hour. Slow and even delivery further improves water use efficiency without incurring the expense and complexity of pulsed delivery equipment.

An emitting pipe is a type of drip irrigation tubing with emitters pre-installed at the factory with specific distance and flow per hour as per crop distance.

An emitter restricts water flow passage through it, thus creating head loss required (to the extent of atmospheric pressure) in order to emit water in the form of droplets. This head loss is achieved by friction / turbulence within the emitter.

Advantages and Disadvantages

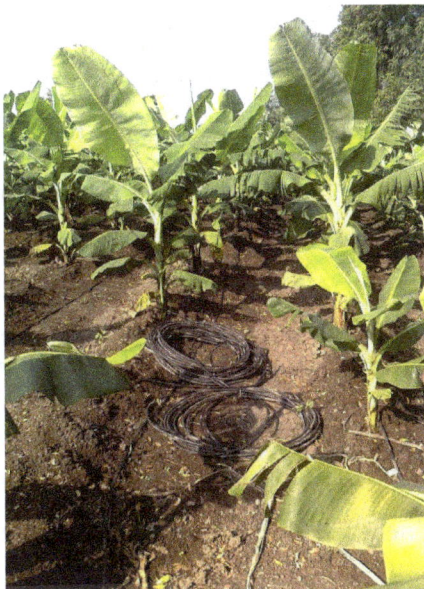

Drip irrigation and spare drip irrigation tubes in banana farm at Chinawal, India

Pot irrigation by On-line drippers

Pressure compensated integral dripper on soilless growing channels

The advantages of drip irrigation are:

- Fertilizer and nutrient loss is minimized due to localized application and reduced leaching.

- Water application efficiency is high if managed correctly

- Field levelling is not necessary.

- Fields with irregular shapes are easily accommodated.

- Recycled non-potable water can be safely used.

- Moisture within the root zone can be maintained at field capacity.

- Soil type plays less important role in frequency of irrigation.

- Soil erosion is lessened.

- Weed growth is lessened.

- Water distribution is highly uniform, controlled by output of each nozzle.

- Labour cost is less than other irrigation methods.

- Variation in supply can be regulated by regulating the valves and drippers.

- Fertigation can easily be included with minimal waste of fertilizers.

- Foliage remains dry, reducing the risk of disease.

- Usually operated at lower pressure than other types of pressurised irrigation, reducing energy costs.

The disadvantages of drip irrigation are:

- Initial cost can be more than overhead systems.

- The sun can affect the tubes used for drip irrigation, shortening their usable life. (This article does not include a discussion of the effects of degrading plastic on the soil content and subsequent effect on food crops. With many types of plastic, when the sun degrades the plastic, causing it to become brittle, the estrogenic chemicals (that is, chemicals replicating female hormones) which would cause the plastic to retain flexibility have been released into the surrounding environment.)

- If the water is not properly filtered and the equipment not properly maintained, it can result in clogging.

- For subsurface drip the irrigator cannot see the water that is applied. This may lead to the farmer either applying too much water (low efficiency) or an insufficient amount of water, this is particularly common for those with less experience with drip irrigation.

- Drip irrigation might be unsatisfactory if herbicides or top dressed fertilizers need sprinkler irrigation for activation.

- Drip tape causes extra cleanup costs after harvest. Users need to plan for drip tape winding, disposal, recycling or reuse.

- Waste of water, time and harvest, if not installed properly. These systems require careful study of all the relevant factors like land topography, soil, water, crop and agro-climatic conditions, and suitability of drip irrigation system and its components.

- In lighter soils subsurface drip may be unable to wet the soil surface for germination. Requires careful consideration of the installation depth.

- most drip systems are designed for high efficiency, meaning little or no leaching fraction. Without sufficient leaching, salts applied with the irrigation water may build up in the root zone, usually at the edge of the wetting pattern. On the other hand, drip irrigation avoids the high capillary potential of traditional surface-applied irrigation, which can draw salt deposits up from deposits below.

- the PVC pipes often suffer from rodent damage, requiring replacement of the entire tube and increasing expenses.

- Drip irrigation systems cannot be used for damage control by night frosts (like in the case of sprinkler irrigation systems)

Uses

Irrigation dripper

Drip irrigation is used in farms, commercial greenhouses, and residential gardeners. Drip irrigation is adopted extensively in areas of acute water scarcity and especially for crops and trees such as coconuts, containerized landscape trees, grapes, bananas, pandey, eggplant, citrus, strawberries, sugarcane, cotton, maize, and potatoes.

Drip irrigation for garden available in drip kits are increasingly popular for the homeowner and consist of a timer, hose and emitter. Hoses that are 4 mm in diameter are used to irrigate flower pots.

Center Pivot Irrigation

Center-pivot irrigation (sometimes called central pivot irrigation), also called waterwheel and circle irrigation, is a method of crop irrigation in which equipment rotates around a pivot and crops are watered with sprinklers. A circular area centered on the pivot is irrigated, often creating a circular pattern in crops when viewed from above (sometimes referred to as *crop circles*). Most center pivots were initially water-powered, and today most are propelled by electric motors.

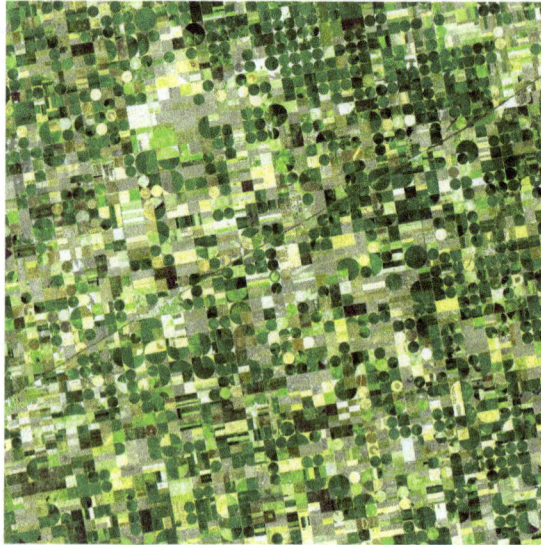

A satellite image of circular fields characteristic of center pivot irrigation, Kansas.

History

Center-pivot irrigation was invented in 1940 by farmer Frank Zybach, who lived in Strasburg, Colorado. It was recognized as a method to improve water distribution to fields.

Overview

Pivot irrigation in progress on a cotton farm.

Center Pivot Irrigation Nozzles

Center pivot irrigation is a form of overhead sprinkler irrigation consisting of several segments of pipe (usually galvanized steel or aluminum) joined together and supported by trusses, mounted on wheeled towers with sprinklers positioned along its length. The machine moves in a circular pattern and is fed with water from the pivot point at the center of the circle. The outside set of wheels sets the master pace for the rotation (typically once every three days). The inner sets of wheels are mounted at hubs between two segments and use angle sensors to detect when the bend at the joint exceeds a certain threshold, and thus, the wheels should be rotated to keep the segments aligned. Center pivots are typically less than 1600 feet (500 meters) in length (circle radius) with the most common size being the standard 1/4 mile (400 m) machine.

Rotator style pivot applicator sprinkler

To achieve uniform application, center pivots require an even emitter flow rate across the radius of the machine. Since the outer-most spans (or towers) travel farther in a given time period than the innermost spans, nozzle sizes are smallest at the inner spans and increase with distance from the pivot point. Aerial views show fields of circles created by the watery tracings of "quarter- or half-mile of the center-pivot irrigation pipe," created by center pivot irrigators which use "hundreds and sometimes thousands of gallons a minute."

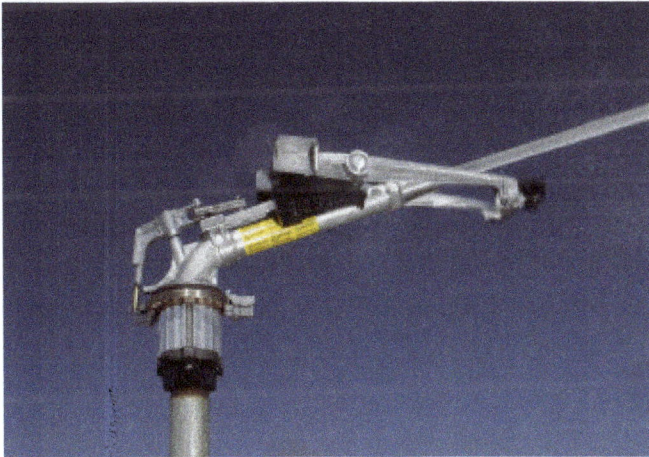

End Gun style pivot applicator sprinkler

Most center pivot systems now have drops hanging from a u-shaped pipe called a *gooseneck* attached at the top of the pipe with sprinkler heads that are positioned a few feet (at most) above the crop, thus limiting evaporative losses and wind drift. There are many different nozzle configurations available including static plate, moving plate and part circle. Pressure regulators are typically installed upstream of each nozzle to ensure each is operating at the correct design pressure.

Drops can also be used with drag hoses or bubblers that deposit the water directly on the ground between crops. This type of system is known as LEPA (Low Energy Precision Application) and is often associated with the construction of small dams along the furrow length (termed furrow diking/dyking). Crops may be planted in straight rows or are sometimes planted in circles to conform to the travel of the irrigation system

Originally, most center pivots were water-powered. These were replaced by hydraulic systems and electric motor-driven systems. Most systems today are driven by an electric motor mounted at each tower.

For a center pivot to be used, the terrain needs to be reasonably flat; but one major advantage of center pivots over alternative systems is the ability to function in undulating country. This advantage has resulted in increased irrigated acreage and water use in some areas. The system is in use, for example, in parts of the United States, Australia, New Zealand, Brazil and also in desert areas such as the Sahara and the Middle East.

Benefits

The center-pivot irrigation system is considered to be a highly efficient system which helps conserve water.

Center pivot irrigation typically uses less water compared to many surface irrigation and furrow irrigation techniques, which reduces the expenditure of and conserves water. It also helps to reduce labor costs compared to some ground irrigation techniques, which are often more labor-intensive. Some ground irrigation techniques involve the digging of channels on the land for the water to flow, whereas the use of center-pivot irrigation can reduce the amount of soil tillage that occurs and helps to reduce water runoff and soil erosion that can occur with ground irrigation. Less tillage encourages more organic materials and crop residue to decompose back into the soil, and reduces soil compaction.

In the United States early settlers of the semiarid High Plains were plagued by crop failures due to cycles of drought, culminating in the disastrous Dust Bowl of the 1930s. It was only after World War II when center pivot irrigation became available that the land mass of the High Plains aquifer system was transformed into one of the most agriculturally productive regions in the world.

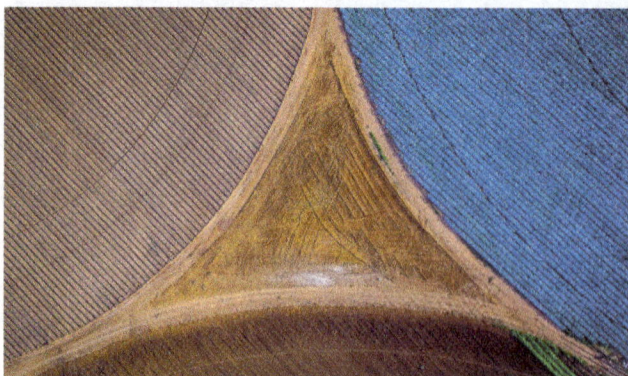

The crops are planted in circles for efficient irrigation.

Risks: Shrinking Irreplaceable Aquifers

Center-pivot irrigation in the Arabian Desert.

Ogallala Aquifer

Groundwater level elevation decreases when the rate of extraction by irrigation exceeds the rate of recharge. At some places, the water table was measured to drop more than five feet (1.5 m) per year at the time of maximum extraction. In extreme cases, the deepening of wells was required to reach the steadily falling water table. In the 21st century, recognition of the significance of the Ogallala Aquifer (also known as the High Plains Aquifer) has led to increased coverage from regional and international journalists.

By 2013 it was shown that as the water consumption efficiency of the center-pivot irrigator improved over the years, farmers planted more intensively, irrigated more land, and grew thirstier crops.

Writer Emily Woodson characterized the increased use of the center pivot irrigation system as part of a profound attitude shift towards modernism (expensive tractors, center-pivot irrigation, dangerous new pesticides) and away from traditional farming that took place in the mid-1970s and 1980s in the United States. A new generation chose high-risk, high-reward crops such as irrigated corn or peanuts, which require large quantities of ground water, fertilizer and chemicals. The new family farm corporations turned many pastures into new cropland and were more interested in rising land prices than water conservation.

A May 2013 *New York Times* article "Wells dry, fertile plains turn to dust" recounts the relentless decline of parts of the High Plains Aquifer System. One of the world's largest aquifers, it covers an area of approximately 174,000 mi² (450,000 km²) in portions of the eight states of South Dakota, Nebraska, Wyoming, Colorado, Kansas, Oklahoma, New Mexico, and Texas, beneath the Great Plains in the United States.

In parts of the United States, sixty years of the profitable business of intensive farming using huge center-pivot irrigators has emptied parts of the High Plains Aquifer. It would take hundreds to thousands of years of rainfall to replace the groundwater in the dried up aquifer. In 1950, irrigated cropland covered 250,000 acres. With the use of center-pivot irrigation, nearly three million acres of land were irrigated in Kansas alone. In some places in the Texas Panhandle, the water table has been drained (dewatered). "Vast stretches of Texas farmland lying over the aquifer no longer support irrigation. In west-central Kansas, up to a fifth of the irrigated farmland along a 100-mile (160 km) swath of the aquifer has already gone dry."

Linear/lateral Move Irrigation Machines

A small center pivot system from beginning to end.

The above-mentioned equipment can also be configured to move in a straight line where it is termed a *linear move, lateral move, wheelmove* or *side-roll* irrigation system. In this case the water is supplied by an irrigation channel running the length of the field and positioned either at one side or midway across the field width. The motor and pump equipment is mounted on a cart adjacent to the supply channel that travels with the machine.

Farmers may opt for linear moves to conform to existing rectangular field designs such as those converting from furrow irrigation. Lateral moves are far less common, rely on more complex guidance systems, and require additional management compared to center pivot systems. Lateral moves are common in Australia and typically range between 500 and 1,000 meters in length.

Deficit Irrigation

Deficit irrigation (DI) is a watering strategy that can be applied by different types of irrigation application methods. The correct application of DI requires thorough understanding of the yield response to water (crop sensitivity to drought stress) and of the economic impact of reductions in harvest. In regions where water resources are restrictive it can be more profitable for a farmer to maximize crop water productivity instead of maximizing the harvest per unit land. The saved water can be used for other purposes or to irrigate extra units of land. DI is sometimes referred to as incomplete supplemental irrigation or regulated DI.

Definition

Deficit irrigation (DI) has been reviewed and defined as follows:

"Deficit irrigation is an optimization strategy in which irrigation is applied during drought-sensitive growth stages of a crop. Outside these periods, irrigation is limited or even unnecessary if rainfall provides a minimum supply of water. Water restriction is limited to drought-tolerant phenological stages, often the vegetative stages and the late ripening period. Total irrigation application is therefore not proportional to irrigation requirements throughout the crop cycle. While this inevitably results in plant drought stress and consequently in production loss, DI maximizes irrigation water productivity, which is the main limiting factor (English, 1990). In other words, DI aims at stabilizing yields and at obtaining maximum crop water productivity rather than maximum yields (Zhang and Oweis, 1999)."

Crop Water Productivity

Crop water productivity (WP) or water use efficiency (WUE) expressed in kg/m^3 is an efficiency term, expressing the amount of marketable product (e.g. kilograms of grain) in relation to the amount of input needed to produce that output (cubic meters of water). The water used for crop production is referred to as crop evapotranspiration. This is a combination of water lost by evaporation from the soil surface and transpiration by the plant, occurring simultaneously. Except by modeling, distinguishing between the two processes is difficult. Representative values of WUE for cereals at field level, expressed with evapotranspiration in the denominator, can vary between 0.10 and 4 kg/m3.

Experiences with Deficit Irrigation

For certain crops, experiments confirm that DI can increase water use efficiency without severe yield reductions. For example for winter wheat in Turkey, planned DI increased yields by 65% as compared to winter wheat under rainfed cultivation, and had double the water use efficiency as compared to rainfed and fully irrigated winter wheat. Similar positive results have been described for cotton. Experiments in Turkey and India indicated that the irrigation water use for cotton could be reduced to up to 60 percent of the total crop water requirement with limited yield losses. In this way, high water productivity and a better nutrient-water balance was obtained.

Certain Underutilized and horticultural crops also respond favorably to DI, such as tested at experimental and farmer level for the crop quinoa. Yields could be stabilized at around 1.6 tons per hectare by supplementing irrigation water if rainwater was lacking during the plant establishment and reproductive stages. Applying irrigation water throughout the whole season (full irrigation) reduced the water productivity. Also in viticulture and fruit tree cultivation, DI is practiced.

Scientists affiliated with the Agricultural Research Service (ARS) of the USDA found that conserving water by forcing drought (or deficit irrigation) on peanut plants early in the growing season has shown to cause early maturation of the plant yet still maintain sufficient yield of the crop. Inducing drought through deficit irrigation earlier in the season caused the peanut plants to physiologically "learn" how to adapt to a stressful drought environment, making the plants better able to cope with drought that commonly occurs later in the growing season. Deficit irrigation is beneficial for the farmers because it reduces the cost of water and prevents a loss of crop yield (for certain crops) later on in the growing season due to drought. In addition to these findings, ARS scientists suggest that deficit irrigation accompanied with conservation tillage would greatly reduce the peanut crop water requirement.

For other crops, the application of deficit irrigation will result in a lower water use efficiency and yield. This is the case when crops are sensitive to drought stress throughout the complete season, such as maize.

Apart from university research groups and farmers associations, international organizations such as FAO, ICARDA, IWMI and the CGIAR Challenge Program on Water and Food are studying DI.

Reasons for Increased Water Productivity Under Deficit Irrigation

If crops have certain phenological phases in which they are tolerant to water stress, DI can increase the ratio of yield over crop water consumption (evapotranspiration) by either reducing the water loss by unproductive evaporation, and/or by increasing the proportion of marketable yield to the totally produced biomass (harvest index), and/or by increasing the proportion of total biomass production to transpiration due to hardening of the crop - although this effect is very limited due to the conservative relation between biomass production and crop transpiration, - and/or due to adequate fertilizer application and/or by avoiding bad agronomic conditions during crop growth, such as water logging in the root zone, pests and diseases, etc.

Advantages

The correct application of DI for a certain crop:

- maximizes the productivity of water, generally with adequate harvest quality;

- allows economic planning and stable income due to a stabilization of the harvest in comparison with rainfed cultivation;

- decreases the risk of certain diseases linked to high humidity (e.g. fungi) in comparison with full irrigation;

- reduces nutrient loss by leaching of the root zone, which results in better groundwater quality and lower fertilizer needs as for cultivation under full irrigation;

- improves control over the sowing date and length of the growing period independent from the onset of the rainy season and therefore improves agricultural planning.

Constraints

A number of constraints apply to deficit irrigation:

- Exact knowledge of the crop response to water stress is imperative.

- There should be sufficient flexibility in access to water during periods of high demand (drought sensitive stages of a crop).

- A minimum quantity of water should be guaranteed for the crop, below which DI has no significant beneficial effect.

- An individual farmer should consider the benefit for the total water users community (extra land can be irrigated with the saved water), when he faces a below-maximum yield;

- Because irrigation is applied more efficiently, the risk for soil salinization is higher under DI as compared to full irrigation.

Modeling

Field experimentation is necessary for correct application of DI for a particular crop in a particular region. In addition, simulation of the soil water balance and related crop growth (crop water productivity modeling) can be a valuable decision support tool. By conjunctively simulating the effects of different influencing factors (climate, soil, management, crop characteristics) on crop production, models allow to (1) better understand the mechanism behind improved water use efficiency, to (2) schedule the necessary irrigation applications during the drought sensitive crop growth stages, considering the possible variability in climate, to (3) test DI strategies of specific crops in new regions, and to (4) investigate the effects of future climate scenarios or scenarios of altered management practices on crop production.

References

- Richard Bulliet, Pamela Kyle Crossley, Daniel Headrick, Steven Hirsch. Pages 53-56 (2008-06-18). The Earth and Its Peoples, Volume I: A Global History, to 1550. Books.google.com. ISBN 0618992383. Retrieved 2012-06-19.

- Frenken, K. (2005). Irrigation in Africa in figures – AQUASTAT Survey – 2005 (PDF). Food and Agriculture Organization of the United Nations. ISBN 92-5-105414-2. Retrieved 2007-03-14.

- Drainage Manual: A Guide to Integrating Plant, Soil, and Water Relationships for Drainage of Irrigated Lands. Interior Dept., Bureau of Reclamation. 1993. ISBN 0-16-061623-9.

- Drainage Manual: A Guide to Integrating Plant, Soil, and Water Relationships for Drainage of Irrigated Lands. Interior Dept., Bureau of Reclamation. 1993. ISBN 0-16-061623-9.

- R. Goyal, Megh (2012). Management of drip/trickle or micro irrigation. Oakville, CA: Apple Academic Press. p. 104. ISBN 978-1926895123.

- Morris, John Miller (2003). Sherry L. Smith, ed. The Future of the Southern Plains. Norman, Oklahoma: University of Oklahoma Press. p. 275. ISBN 0806137355.

- "Polyester ropes natural irrigation technique". Entheogen.com. Archived from the original on April 12, 2012. Retrieved 2012-06-19.

- Mader, Shelli (May 25, 2010). "Center pivot irrigation revolutionizes agriculture". The Fence Post Magazine. Retrieved June 6, 2012.

- Alfred, Randy (July 22, 2008). "July 22, 1952: Genuine Crop-Circle Maker Patented". Wired Magazine. Retrieved June 6, 2012.

- "Growing Rice Where it has Never Grown Before: A Missouri research program may help better feed an increasingly hungry world". College of Agriculture, Food and Natural Resources, University of Missouri. July 3, 2008. Retrieved June 6, 2012.

- Evans, R.O.; et al. (March 1997). "Center Pivot and Linear Move Irrigation System" (PDF). North Carolina Cooperative Extension Service, North Carolina State University. Retrieved June 6, 2012.

- "Africa, Emerging Civilizations In Sub-Sahara Africa. Various Authors; Edited By: R. A. Guisepi". History-world.org. Retrieved 2012-06-19.

- Mader, Shelli (May 25, 2010). "Center pivot irrigation revolutionizes agriculture". The Fence Post Magazine. Retrieved June 6, 2012.

Permissions

All chapters in this book are published with permission under the Creative Commons Attribution Share Alike License or equivalent. Every chapter published in this book has been scrutinized by our experts. Their significance has been extensively debated. The topics covered herein carry significant information for a comprehensive understanding. They may even be implemented as practical applications or may be referred to as a beginning point for further studies.

We would like to thank the editorial team for lending their expertise to make the book truly unique. They have played a crucial role in the development of this book. Without their invaluable contributions this book wouldn't have been possible. They have made vital efforts to compile up to date information on the varied aspects of this subject to make this book a valuable addition to the collection of many professionals and students.

This book was conceptualized with the vision of imparting up-to-date and integrated information in this field. To ensure the same, a matchless editorial board was set up. Every individual on the board went through rigorous rounds of assessment to prove their worth. After which they invested a large part of their time researching and compiling the most relevant data for our readers.

The editorial board has been involved in producing this book since its inception. They have spent rigorous hours researching and exploring the diverse topics which have resulted in the successful publishing of this book. They have passed on their knowledge of decades through this book. To expedite this challenging task, the publisher supported the team at every step. A small team of assistant editors was also appointed to further simplify the editing procedure and attain best results for the readers.

Apart from the editorial board, the designing team has also invested a significant amount of their time in understanding the subject and creating the most relevant covers. They scrutinized every image to scout for the most suitable representation of the subject and create an appropriate cover for the book.

The publishing team has been an ardent support to the editorial, designing and production team. Their endless efforts to recruit the best for this project, has resulted in the accomplishment of this book. They are a veteran in the field of academics and their pool of knowledge is as vast as their experience in printing. Their expertise and guidance has proved useful at every step. Their uncompromising quality standards have made this book an exceptional effort. Their encouragement from time to time has been an inspiration for everyone.

The publisher and the editorial board hope that this book will prove to be a valuable piece of knowledge for students, practitioners and scholars across the globe.

Index

www.ingramcontent.com/pod-product-compliance
Lightning Source LLC
Chambersburg PA
CBHW061303190326
41458CB00011B/3751